Chicken

Animal
Series editor: Jonathan Burt

Already published

Chicken

Annie Potts

REAKTION BOOKS

In loving memory of Faith Potts (1933–2011)

Published by
REAKTION BOOKS LTD
33 Great Sutton Street
London EC1V 0DX, UK
www.reaktionbooks.co.uk

First published 2012
Copyright © Annie Potts 2012

Printed and bound in China by C&C Offset Printing Co., Ltd

British Library Cataloguing in Publication Data
Potts, Annie. 1965–
    Chicken. – (Animal)
    1. Chickens—Evolution 2. Chickens—Behavior.
    3. Chickens—Folklore.   4. Chickens in art.
    I. Title II. Series
    598.6'25-DC22

ISBN 978 1 86189 858 6

Contents

1 From *T. rex* to Transylvanian Naked Necks

The discovery in 1861 of the fossilized remains of the archaeopteryx first led palaeontologists to speculate that birds evolved from dinosaurs. The archaeopteryx, thought to have lived in Jurassic Bavaria around 147 million years ago, possessed rudimentary feathers and wings. We now know that other dinosaurs also grew feathers; some, like modern-day birds, were warm-blooded and hollow-boned. Advances in genetic research suggest that the ancestry of birds is much more closely and complexly connected to the carnivorous theropods than any other dinosaurs.[1] The humble chicken may in fact be the nearest living relative of the largest and most notorious predatory dinosaur, *Tyrannosaurus rex*: in 2007 protein traces recovered from the soft tissue of a 68-million-year-old *T. rex* femur bone were observed to match most closely those of the *Gallus* (chicken) species.[2]

The direct ancestor of the chicken (the 'pre-chicken') evolved somewhat later, probably around 50 million years ago. Fossils of *Gallus* are relatively scarce and have mainly been found in southern Europe. Fossil discoveries here imply that the ancestors of today's jungle fowl dispersed into warmer territories during periods of glaciation; when temperatures increased and glaciers

Hatty Morris, *Chicken and T-Rex*, 2009, charcoal on paper.

7

retreated these subpopulations spread north and eventually reunited – the length of time apart resulting in the evolution of different species or subspecies. Evidence of prehistoric *Gallus* species surviving into historic Europe has not been found, possibly because these birds died out during subsequent glaciation or were made extinct by predators (including humans). Those species of jungle fowl we know today are therefore probably descended from prehistoric populations that took refuge during glaciation in what is now the region of South-East Asia. They would have been prevented from travelling north by the Himalayas, therefore evolving in relative isolation. Other *Gallus* subpopulations may have evolved in the Far East: there are archaeological signs of domesticated chickens in China several thousand years before their domestication in India. Moreover, domestic chicken breeds in Asia are sufficiently different from others to indicate separate evolutionary backgrounds.[3]

JUNGLE FOWL

The wild jungle fowl, from which the domestic chicken descends, belongs to the order Galliformes, suborder Galli, family Phasianidae (which also includes pheasants, partridges, peafowl, quail, francolins and monals). Subfamilies of Phasianidae are distinguished according to the manner of tail moulting. Chickens belong to the second subfamily (Phasianinae), in which moulting is centripetal (the pattern of tail-feather loss occurs from the outside to the centre).

Phasianids are generally territorial ground-dwelling non-migratory birds. They tend to have shortish wings, solid legs and short, strong beaks. Males are larger than females, possess spurs, and have brightly coloured plumage and facial ornaments such as wattles and combs. Hens are often described as

drab in comparison, but their less eye-catching plumage cam-
ouflages them when they are nesting on the ground to incubate
eggs. These birds are reasonably powerful short-range flyers,
which allows speedier escape from predators and safer roosting
in trees at dusk.

The four recognized species of modern wild jungle fowl
include the red (*Gallus gallus*), the Sri Lanka (*Gallus g. lafayetii*),
the green (*Gallus g. varius*) and the grey (*Gallus g. sonneratii*). Red,
grey and Sri Lanka jungle fowl prefer tropical or subtropical
forest habitats, while the green tends to live near the seashore
or by scrubland bordering cultivated land. The grey inhabits
western and southern India, while the Sri Lanka lives naturally
only where its name suggests. The unusual green jungle fowl,
found in Java and islands eastward, where it occupies different
habitats from the red, is thought to be the most primitive of the
four, boasting sixteen tail feathers and short hackle feathers,
whereas the other species possess fourteen tail feathers and
long pointed hackle feathers.

GALLUS VARIUS.

There are no subspecies of grey, green or Sri Lanka jungle fowl, but several subspecies of red jungle fowl exist: the Cochin-Chinese red (*G. g. gallus*), located in Thailand, Laos, Cambodia and south Vietnam; the Indian red (*G. g. murghi*) of north and north-east India, Nepal and Bangladesh; the Burmese (*G. g. spadiceus*), found in south-west China, Myanmar, Malaysia, north Sumatra and Thailand (except in the east); the Javan red (*G. g. bankiva*), located in south Sumatra, Bali and Java; and the Tonkinese red (*G. g. jabouillei*) of south-west China and north Vietnam.[4] These five distinct subspecies of red jungle fowl are unable to interbreed with each other but they are capable of successfully mating with domestic chickens.

Wild and domesticated red jungle fowl were introduced to Kenya, the Philippines and other Pacific islands in Micronesia, Melanesia and Polynesia (forced introductions to Australia, New Zealand and North America were not successful). Red jungle fowl

Gallus gallus varius, the green jungle fowl; a hand-coloured lithograph from Daniel Giraud Elliot, *A Monograph of the Phasianidae (Family of Pheasants)* (1872).

A wild rooster wandering among tourists on Poipu Beach, Kauai, Hawai'i.

were also brought to Hawai'i around 700 years ago by colonizing Polynesians; today families of feral chickens can be seen roaming the streets, forests and even the beaches of Kauai.

Domestication of the chicken is thought to have occurred around 8,000 to 10,000 years ago.[5] Charles Darwin believed that the red jungle fowl of South-East Asia was the direct and single ancestor of the domestic chicken; but while modern genetic research verifies this species' important role as a progenitor of domestic chickens (to the extent that *G. g. domesticus* is recognized as a subspecies of red jungle fowl, rather than a separate species),[6] debate continues as to the involvement of multiple hereditary and geographical origins.[7] Some scientists hypothesize several domestication events occurring independently across various regions in Southern Asia and involving inter-species hybridization within the genus *Gallus*. The theory of multiple progenitors is supported by the yellow leg and skin colour of many domestic chickens, which depends on a gene deriving not from the red jungle fowl but from the grey.[8]

The earliest remains of domestic chickens (distinguished from wild fowl remains by their larger skeletal size) are at least 7,500 years old and have been found in north-east China in sixteen Neolithic sites along Huang He (Yellow River), and in the Indus Valley in Pakistan. Chickens were most likely first taken north from their origins in South-East Asia to China, where they became established by 6000 BC. They were then transported eastwards, as shown by evidence of their presence in Russia, Turkey and eastern Europe around 3000 BC. The global transportation of chickens continued into Spain by 1200 BC and then to north-west Europe by 500 BC: the earliest records of domestic chickens in Britain can be dated to around 55 BC. It is presumed that chickens were introduced to North America around 1,600 years later.[9]

As early as 1500 BC depictions of chickens emerge in Egyptian hieroglyphic art. A rooster is visible in a scene in the tomb of Rekhmara (*c.* 1500 BC), and chicken symbolism has also been identified in Tutankhamen's tomb (*c.* 1400 BC). Interestingly, chickens then disappear from North African graphic records until around 650 BC. Fourth-century Greek records state that Egyptians had mastered poultry husbandry and had been practising artificial incubation of chicks for many years. The giant incubators invented by the ancient Egyptians – possibly to sustain the masses of labourers involved in the construction of pyramids – were capable of hatching up to 15,000 eggs at one time. The eggs were warmed by fires and turned at regular intervals by attendants. The ancient Chinese also created incubators (using fires or rotting manure), perhaps to provide eggs for the builders of the Great Wall of China.[10] Thus behind our most wondrous human monuments, it seems, there roosts the chicken.

In sub-Saharan Africa the earliest chicken remains – found in Mali, Nubia, the East African Coast and South Africa – have been dated to around AD 500, leading archaeologists to speculate that chickens came to Africa via the Nile Valley, or through early Graeco-Roman east coast trade.[11]

Similarly, the spread of jungle fowl from South-East Asia to the Mediterranean may have occurred via commercial (or military) contacts. The Indian practices of cockfighting and chicken husbandry (for meat and eggs) were probably passed on to the Persian invaders in 400 BC, who in turn transferred them to the Romans and Greeks. The Romans practised caponizing and force-feeding of chickens, and also knew about animal husbandry principles of 'hybrid vigor [and] sperm competition'.[12] They produced two

dual-purpose breeds for entertainment (cockfighting) and food, and two other breeds used solely for competitive fighting.

The islands of the Pacific also have an ancient history involving chickens. In the eighteenth century James Cook noticed chickens on the Easter Islands, New Caledonia and elsewhere in the Pacific. The Tahitians claimed chickens were made at the same time as humans by the god Taarva and had always lived amongst them. Hawai'ians told Cook the legend of a gigantic bird who, upon settling on the water, laid an egg that opened to produce the island of Hawai'i.

Debate continues over whether or not chickens were present in South America before the arrival of Columbus and the introduction of Spanish chickens in the sixteenth century. If they were, this implies very early contact between Polynesian or Asian peoples and the indigenous South Americans. The grouse-like appearance of the Araucana chicken of South America (unique among chicken breeds for the hens' blue and green eggs) may be the result of interbreeding between native grouse and the domestic fowl brought to the Americas by visitors whose travels are unrecorded in existing historical annals.[13] Intriguingly, both

The Araucana hen, native to South America, lays unique green and blue eggs.

George Stubbs, 'Fowl, Lateral View with Most Feathers Removed' and 'Fowl Skeleton – Lateral View', 'finished studies' for *A Comparative Anatomical Exposition of the Structure of the Human Body with that of a Tiger and a Common Fowl* (1794–1806), pencil on paper.

Araucana and Asiatic chickens have red earlobes, whereas those of Mediterranean breeds are white.

Those mammal and bird species that have been successfully domesticated tend to share certain biological features and behaviours that facilitate their control by humans. They form large groups in which a social order is important, making it easier to manage them in substantial numbers. They are able to reproduce in captivity. Parent–offspring relationships are important, often involving precocial development of the young and critical periods of bonding between a mother and her brood. This makes these creatures easy to tame since they bond readily to humans if exposed to people soon after birth; it also facilitates the use of surrogate parents. Their ability to modify foraging and other behaviours according to different climactic conditions is also usually well developed.[14] Chickens, to their credit and to their disadvantage, demonstrate all these characteristics.

Four historical stages have been identified in the 'taming' of the chicken for human needs. Initial domestication was in the interests of religion and culture as chickens increasingly became a focus of folklore, mythology and religious symbolism. Birds were selectively bred to produce specific physical traits (particular colours and/or morphological features). Feathers, bones and eggs became culturally significant and were used as tools or for decoration.[15] Roosters were valued for their fighting and time-keeping abilities, their crowing heralding a new day. In this early period in the development of the domestic chicken, consumption of birds was not routinely practised; in some societies the eating of chickens was strictly forbidden, although some may have been slaughtered for sacrificial rituals.

The second stage of domestication corresponded with the geographical dispersal of chickens away from their origins, resulting in the eventual creation of distinct regional breeds. Cockfighting,

Cockfighting mosaic uncovered in the House of the Labyrinth, Pompeii, c. AD 1.

extremely popular in ancient Rome, Greece, Germany and some regions of Asia, had a huge impact on the domestication and forced diaspora of chickens during this phase.

The third stage, associated with the nineteenth-century 'hen craze', coincided with the proliferation of industry and agriculture in Europe and North America, and involved an avid popular interest in the selective mating of certain breeds. Emphasis was placed less on the production of eggs or meat and more on the selective breeding of superlative birds for exhibition – although the main breeds used by the poultry industry today, such as the Cornish, White Leghorn and Plymouth Rock, emerged during the hen craze.

The fourth stage coincided with the twentieth century's development and expansion of industrialized farming and poultry capitalism.[16] Burgeoning scientific knowledge about poultry genetics and breeding, nutrition and disease, as well as innovations in chicken housing and farm technologies, sealed the fates of the contemporary layer hen and broiler chick.

Domestication has also been associated with the formation of specific terminologies to describe the age and sex of chickens. In the UK, Ireland and Canada, adult male fowl are called cocks, while they are more likely to be referred to as roosters in America and Australasia. Cockerel is the term given to a male under one year of age, and a castrated cockerel is a capon (the removal of testicles is said to produce more tender meat on male birds). A female chicken is a pullet up to one year of age, when she is called a hen.

ON SHOW: FANCY BREEDS

For many centuries the selective breeding of chickens was not practised with any gusto except within cockfighting fraternities.

COMPOSED BY

FRANCIS H. BROWN

Poultry shows were all the rage in Britain and North America during the late 19th century.

When cockfighting was outlawed as a recreational 'sport' in Britain in 1849 (under Queen Victoria's directive), the 'poultry show' emerged as an alternative, more acceptable form of competition. Attention turned to defining and enhancing specific varieties of chickens. The first British and American exhibitions

THE CONSTITUTIONAL WALK.

Lady. "DEAR, DEAR, IT'S COMING ON TO RAIN! RUN, JAMES! QUICK, AND FETCH AN UMBRELLA, AND TWO PARASOLS. I'M AFRAID MY POOR DEAR COCHINS WILL GET THE RHEUMATISM."

of 'authentic' chicken breeds took place at this time, marking the era of the hen craze. In the early nineteenth century only a handful of chicken breeds – such as the Old English Game, Dorking and Hamburg – were formally recognized in Britain; by the start of the twentieth century scores of new breeds had been created and Breed Societies had been established, while poultry shows had become popular public events. The world's first book of poultry standards was released in 1865 in Britain, although it was not until 1877 that The Poultry Club of Great Britain was established.

The American Poultry Association (APA) held its inaugural meeting in Buffalo, New York, in 1873. Its purpose was 'to standardize the varieties of domestic fowl so that a fair decision could

Comb types:
1 single
2 pea
3 cushion
4 rose
5 v-shaped
6 buttercup
7 strawberry
8 walnut

be made as to which qualities marked prize winners'.[17] In 1874 the *American Standard of Excellence* was drafted. The later form of this text, the *American Standard of Perfection*, divides chickens into two size categories, large fowls and bantams, and lists the ideal physical and temperamental characteristics for all recognized breeds. Classes of large breeds now include American, Asiatic, English, Mediterranean, Continental and All Other Standard breeds. Bantam subcategories include miniature versions of large breeds – Game, Single Comb Clean Legged Other Than Game, Rose Comb Clean Legged, All Other Combs Clean Legged and Feather Legged – and those naturally diminutive breeds classified as True Bantams.

A breed is a group within a species that shares 'definable and identifiable characteristics (visual, performance, geographical, and/or cultural) which allow it to be distinguished from other groups within the same species'.[18] Within breeds different types and varieties also exist. 'Type' refers to the breed's primary purpose, whether as an exhibition or 'utility' type; chickens may be grouped within Light Breed, Heavy Breed, Hard Feather, Soft Feather and Rare Breed categories. Light Breeds tend to be known for their egg-laying abilities, while Heavy Breeds have been developed primarily for the qualities of their flesh. Hard Feather breeds are the game birds historically used for cockfighting, while Soft Feather refers to breeds such as the Silkie and Brahma. Rare Breeds (such as the Ancona and Modern Game) are those associated with specialist clubs supporting their survival.[19] 'Varieties' are defined by features such as plumage (more than 30 genes influence feather colours and patterns in chickens), comb (shape, size and colour), tail (shape, size and angle), wattles, earlobes, eggshell colour, presence of beards or muffs, and number of toes (most chickens have four on each foot but some have five).

Contemporary breeds derive predominantly from Mediterranean and Asiatic types. Mediterranean breeds are small-bodied, small-boned and recognizable by their white earlobes. They are reputed to be noisy, excitable birds, good layers but poor brooders, this latter characteristic contributing to the current critical conservation status of the Andalusian, Catalana and Sicilian Buttercup breeds. Asiatic breeds (Langshans, Brahmas and Cochins) are gentle, large-bodied, heavy-boned and fleshy, and tend to possess red earlobes. Langshans hark from the district of that name north of the Yangtze River. Brahmas, named after the Brahmaputra River in India but thought to have been produced in America during the 1840s from breeds imported from India (the Chittagong) and China (the Shanghai), are 'cuddly giants' with feathered legs and colourful plumage.[20]

The Arms for Dorking Urban District Council: motto 'Virtute et Vigilantia' – by courage and vigilance. The Dorking breed (known for its five-toed feet) has been in existence for around 2,000 years and was possibly brought to Britain during the Roman invasion in the first century BC.

The Malay is the tallest breed of chicken, standing at three feet (90 cm).

The spectacular Poland.

When Julius Caesar invaded England in 55 BC, he found that chickens were already being bred there for fighting purposes.[21] Old English Game birds are an ancient breed of chicken from which two contemporary types have evolved: the Carlisle and Oxford. The Indian Game bird, also originating in Britain, is thought to result from the cross-breeding of Malay, Asil (an ancient Indian breed) and Old English Game. The Indian Game has in turn been selectively cross-bred with Dorking and Sussex to produce large, heavy offspring for the chicken meat industry. The Modern Game, also originally developed for cockfighting, never actually entered a pit; these birds are now bred for show purposes only.[22]

Like the Dorking, Sussex and Cornish breeds of Britain, the names of several American breeds reveal their geographical affiliations. The Rhode Island Red, created as a 'dual-purpose utility bird' by mixing indigenous fowl with imported chickens from the Far East (including Shanghais, Malays and Leghorns), was initially shown in Massachusetts in 1880.[23] Two types of Rhode Island Red exist, determined by comb shape (the Single Comb and the Rose Comb). The less well-known Rhode Island White was first recognized in the *Standard* in 1922. The Delaware, another hardy American breed originating in 1940, is thought to be the outcome of interbreeding Barred Plymouth Rock roosters with New Hampshire hens. From 1940 to 1960 this bird was the dominant breed in the broiler industry of Delmarva Peninsula (until it was replaced by the Cornish-Plymouth Rock cross). Now its conservation status is considered critical.[24] Another robust breed from America, developed primarily for eggs and meat, is the Wyandotte, whose name derives from the Wendat indigenous peoples of upstate New York and Ontario.[25]

Of the Continental breeds, perhaps the most popular – based on appearance – is the Polish or Poland. Possibly deriving from

The Silver-Laced Wyandotte originated in New York in 1865, possibly the result of crossing a Spangled Hamburgh rooster with either a Cochin or Brahma hen (the Sebright bantam might also have entered the mix).

Transylvanian
Naked Neck.

The Yokohama is
classified as a rare
breed.

the Netherlands or Italy, Polish chickens are known as 'top-knots' due to their wild crest feathers; some varieties also sport beards and muffs. Polands possess a v-shaped (horn) comb, an attribute shared by the Houdan, which is a French breed, atypically five-toed (like the Dorking), whose conservation status is listed as critical.[26]

The large sub-classification known as 'All Other' boasts a wide variety of different breeds, such as Araucanas (from South America), Cubalayas (a rare Cuban breed with ancestral ties to the Philippines), Malays and Shamos. Some of the breeds in this class are striking in appearance, like the Frizzles, Naked Necks and Sultans. Frizzles have been bred for their aesthetic and exhibition appeal: the feathers curl towards the bird's head, resembling a frizzy hairstyle, and the comb, wattles and earlobes are bright red. In sharp contrast, Naked Necks (also known as Turkens or Transylvanian Naked Necks) have, as their name suggests, no feathers on the neck and sometimes just sparse plumage around the head. Ostensibly 'freaks of nature', they may have resulted from the long-ago mating of a chicken with a turkey.[27] Sultans, named 'Serai-Tavuk' (fowls of the sultan) in their native Turkey, were kept as pets and 'living garden ornaments' by Turkish royalty for centuries.[28] This breed is small with bright red v-shaped combs, five-toed feet and wildly feathered crests, as well as beards, muffs, 'whiskers' and feathered legs.[29]

Another breed popular because of its unusual appearance – as well as its gentle, friendly nature – is the ancient Silkie, which comes in both large and bantam form. Marco Polo possibly observed Silkies during his thirteenth-century travels through China, because he recorded an encounter with a 'furry' chicken unable to fly. The five-toed Silkies possess soft, light plumage and a melanotic gene resulting in blue earlobes and bluish-black skin, beaks, bones and flesh.[30]

Chinese Silkie.

While most bantam breeds are miniature versions of large fowl, several breeds of chicken exist only in diminutive form. These are the 'True Bantams', including Pekin, Rosecomb and Sebright breeds. Pekins (originally from China) have been likened to balls of fluff on legs; even the feet of this breed are feathered. Rosecombs, present in Britain for at least 600 years and declared

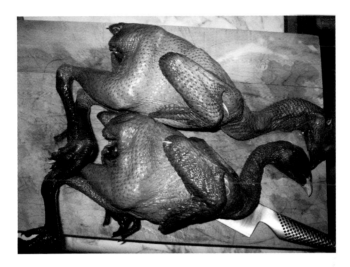

The dark flesh of the Chinese Silkie is considered a delicacy in China and some fusion restaurants in Western countries now serve Silkie meat.

by some to be a British breed, have white earlobes, short beaks and rose-type combs. Sebrights, recognized by their remarkable laced markings, were developed more than 200 years ago by Sir John Sebright, possibly by crossing a Rosecomb bantam with Polish, Nankin and Hamburgh varieties.[31]

Paradoxically, although they exist in the billions, layer hens and broiler (or meat) chicks are the breeds of *Gallus* least on show; that is, until they appear on supermarket shelves or in cans of pet food. Known as 'commercial hybrids', they are the result of repeated intensive selective cross-breeding to produce prolific layers or fast-growing chicks. These 'utility' breeds are the focus of chapter Six.

2 Chicken Wisdom

While studies on 'intelligence' in birds have focused mainly on parrots, corvids and songbirds, substantial research also exists in relation to the distinctive perceptual, cognitive, social and emotional lives of chickens. Despite overwhelming evidence to the contrary, there remains, in Western culture at least, a tendency to associate chickens with stupidity. This has not always been the case: in fact, the term 'bird-brain' – meaning dimwitted or foolish (and applied to certain other birds as well as chickens) – was first coined as recently as the 1920s; prior to this, in both ancient and more recent societies, roosters and hens were respected and admired for their abilities to guard, protect, nurture and communicate with each other.

The prevalent assumption that chickens are unintelligent can be traced to two main biases. We tend to perceive species similar in appearance to us as being more 'mindful' and aware; the greater the physical difference between humans and other creatures, the more likely we are to view those beings as driven merely by instinct or impulse.[1] We are also influenced by what is known as the 'small-brain fallacy', the assumption that 'tiny things just can't be intelligent or aware'.[2] This prejudice relies in turn on accepting that the cerebral cortex – the large, convoluted portion of the human brain – is the site of 'intelligence' (for all

Chickens have around 40 times more retinal nerves than humans. Possessing four kinds of cone cells, they are able to see further into the ultraviolet spectrum than we are.

creatures). Because this anatomical structure is insubstantial in birds (the avian cortex is smooth, and does not possess the folds and bumps – the gyri and sulci – associated with mammalian brains), chickens and many other birds are presumed to be 'simple-minded'.

However, although the brain neuroanatomy of mammals and birds diverged during evolution, the mental development of both classes of animal has actually been convergent: while the cortical cells of mammals developed on the surface of the brain, homologous cells in birds were retained deeper within the cortex. This does not indicate that 'bird brains' are inferior; rather, birds process information in different ways and in other locations than mammals. The avian experts Susan Orosz and Gay Bradshaw go so far as to argue that the

> attributes such as linguistic ability, spatial memory, social reasoning, personality, representation of self, tool manipulation, episodic memory, and vocal learning observed in avian species are considered comparable to those in primates . . . Birds are not just a step ahead of reptiles, nor are they emotionally immature, but closer to being feathered apes.[3]

With respect to chickens, the renowned neuroscientist Lesley Rogers, author of *The Development of Brain and Behaviour in the Chicken*, writes of a 'demand to understand the cognitive abilities of the domestic chicken above all avian species, because this bird is the one we have singled out for intensive farming. *Gallus gallus domesticus* is indeed the avian species most exploited and least respected.'[4]

An exploration of chicken wisdom begins with an overview of growth within the egg. An embryo's development is a complex process governed by sensitive periods of change and growth that determine not only the physical well-being of the chick upon hatching, but also initiate the processes required for immediate survival in the world outside the egg. The first sensation to emerge in embryonic chicks is touch (around the sixth day after conception), followed by sound perception (around days twelve to fourteen), taste (day fourteen), light perception (day eighteen) and smell (day twenty).

Developing embryos start vocalizing around eighteen days after conception. They first make distress calls called 'peeps', possibly in response to changes in temperature or other environmental cues. The hen replies to peeps by making clucking noises and gently turning and pecking the eggs. Around five hours before hatching, embryos emit pleasure calls ('twitters'), which occur about fifteen seconds after the hen has turned the eggs and sat back down on her nest. Brooding-like calls occur most frequently during the actual hatching process and resemble the sounds made by young chicks when they are resting close together and feeling sleepy.[5]

Embryos also communicate with each other. Not long before hatching they produce clicking noises, which stop during hatching but resume for several hours after they have hatched. Frequent clicking (for example, three clicks per second) accelerates hatching time and is also thought to speed up the development of less advanced embryos; conversely, low frequency sounds emitted by embryos shortly before hatching, but prior to the clicking sounds, have the effect of slowing down the growth of more advanced embryos. It is also possible that embryos

The early development of the auditory system assists embryonic chicks to form sound memories that enable them to communicate with the hen and their peers as soon as they hatch, thus ensuring proximity to the brood and increasing their chances of survival.

are stimulated by their own heartbeats, as well as by those of their mother.

The left side of the brain dominates the processing of aural information in chickens. During the final stages of incubation an embryo's head is twisted in such a way that the left ear is close to the thorax while the right ear faces away from the body. Each ear therefore receives different information: the left hears the embryo's own heartbeat and the right hears sounds from outside the egg, including the hen's heartbeat and vocalizations. Information from the right ear proceeds to and is processed primarily by the left hemisphere of the brain (and vice versa); this is referred to as lateralization of the brain. The left side of the embryonic brain is therefore primed mainly to process sounds from the external environment.[6]

The visual system is also lateralized or asymmetrical. Exposure to light speeds up embryonic development in the early stages of incubation, while in the later stages it affects the development of optical anatomy. Specifically, the right eye tends to receive more light stimulation during incubation because it is nearer the wall of the egg, while the left eye, closer to the body of the embryo, is exposed to less light. The right eye becomes the stronger, more focused eye by hatching time, and it takes two to three weeks post-incubation for the nerve pathways of the left eye to catch up.[7]

Throughout incubation there are sensitive periods during which light exposure has a more pronounced effect. Exposure to light for as little as two hours on the nineteenth day of incubation is enough to ensure that the left hemisphere (and therefore the right eye) will lead in the task of determining between grains of food and pebbles, while the right hemisphere will dominate in terms of mating and attacking behaviours.

AFTER HATCHING

In order to leave the eggshell a hatchling will 'pip' or break it with her 'egg tooth', the sharp point projecting from the tip of the upper mandible of the chick's beak (the subsequent loss of this temporary protrusion has given rise to the phrase 'as scarce as a hen's tooth'). At the time of hatching the brain and nervous system are already mature enough to enable chicks to learn very rapidly the skills required to survive. During the first day or so of life hatchlings mostly sleep in the nest by the hen, a behaviour possibly assisted by olfactory imprinting, during incubation, on the smells of the hen and nest. They quickly become more active, developing the ability to stand and coordinate mobility. Although sustained by the yolk sac for a couple of days after hatching, chicks also begin to scratch around and peck – with closed beaks – at small three-dimensional objects on the ground.

Filial imprinting refers to the process by which precocial chicks learn to recognize, follow and respond to a mother hen. Theoretically, visual imprinting might begin as soon as the chick is hatched and any moving object might precipitate an imprinting response; but the gradual progression of coordinated motor abilities allows more time to pass so that imprinting might be 'guided' towards the hen's individual characteristics. Visual imprinting is also helped by the auditory imprinting started during incubation; at a few days of age chicks are able to distinguish

Imprinting results in learning through observation of behaviours such as feeding.

between their mother's clucking and the noises of other hens. The previously learnt sounds of the hen, nest and siblings, combined with the swiftly developing sense of sight, allow the chicks to recognize and stay with the hen who protects them from predators and other dangers.

In the early days, when chicks are learning to recognize their mother, the hen is also coming to know her chicks. It is possible that she distinguishes her brood from others on the basis of colour. Likewise, chicks probably also come to identify and follow their siblings in response to brood colour, thus ensuring the family stays in close proximity. As the appearance of siblings changes over time (for example, plumage develops or darkens), chicks find other ways of detecting members of their familial group.[8] They also begin to respond to the hen's alarm calls and display other appropriate fearful behaviours.

Newly hatched chicks begin exploratory pecking at food items with their beaks closed, before moving on to ingest actual objects. The beak has special receptors that help chickens to make exact tactile judgements: birds use their beaks, like humans use hands, as a means to investigate the environment and to inflict aggression upon others. The beak also works like teeth, as a cutting instrument in the processing of food. Feeding quickly becomes a social behaviour for chicks, who eat more when alongside companions who are also eating, and in response to a call from their mother indicating she has found them a treat. Pecking itself is a pleasurable activity for chickens, often performed even if food is not ingested.[9]

Around five or six days after hatching chicks begin to move further apart from the hen to explore the environment away from the nest. By the tenth day they are confidently leaving her for longer periods and beginning to engage in social interactions such as sparring (which marks the progression of spatial learning) and frolicking (a play activity performed chiefly for fun).[10] Education by the hen now takes a different form. Chicks are taught how to forage, range and perch by their mother.

Under natural conditions, at around seven or eight weeks of age the hen will move at dusk, with her chicks, from a roosting spot on the ground to a position in a tree or bush. Initially, she chooses a perch close to the ground, but once the chicks are imitating her behaviour the hen moves higher eventually to join other perching adults at night.

ADULT CHICKENS: PERCEPTIONS AND COGNITION

The mature chicken has acute hearing and an auditory frequency range of 15–10,000 HZ. Hearing is particularly important for chickens since they spend long periods on the ground and need to receive early warning of impending danger via alarm calls from flock members. Vision in adult chickens is also highly specialized. Both eyes together weigh the same as the entire brain. Whereas humans move both eyes to scan a scene or follow some stimulus, chickens shift their heads while their eyes make very small movements. In this way they observe the same object in individually distinct but repeated 'snapshots'. Whereas human eyes have just one fovea, upon which the object of focus is projected, chickens possess more than one area specialized for high-acuity vision. In fact, they have an elongated horizontal strip of retina in which the photoreceptor cells responsible for processing bright light occur in density, and they may look at objects with either the monocular lateral field or the frontal binocular field, sometimes switching between the two.[11] In this way, while searching for food, chickens can obtain a clear view of morsels immediately in front of them on the ground while simultaneously receiving a panoramic view of the more distant environment in order to detect movement and possible danger.[12] Typically, chickens switch from lateral to frontal viewing as their distance from an object decreases. Because chickens can use each eye independently, they are able to swap between eyes and focus with each eye on a different distance.

As mentioned, chickens demonstrate lateralization of the brain. Differences between the left and right hemispheres of the brain are evident in the ways that chickens view various objects. In general, visual information received by the left eye is processed by the right hemisphere and vice versa. A rooster tilts his head to look up with his left eye when watching for overhead predators, but he uses his right eye (left hemisphere) to discriminate between food and non-food objects, and he will attack when using his left eye (right hemisphere) to view a foe.

Chickens have impressive memories. Facial recognition amongst chickens requires 'head on' inspections at a distance of less than 20–30 cm away,[13] but they are able to memorize more than a hundred other chicken faces and recognize familiar birds after several months of separation.[14] Chickens also remember humans and have been shown to turn away from the countenances of people they dislike.[15]

Chickens can also grasp abstract concepts. Three-day-old chicks are capable of identifying a whole object when part of it is obscured – a feat not accomplished by human babies until four to five months of age.[16] And they can locate completely hidden items, suggesting that mental representations of the locations of objects are created and stored from a very early age.[17]

They are also able to anticipate the future. In one study hens were taught to peck coloured buttons for food rewards. If a hen waited a short time before pecking the button, she received a small portion of food; if she delayed pecking the button for longer, she received a food 'jackpot'. Ninety per cent of the time hens chose to wait the extra time so that they might obtain more to eat.[18] The ability to consider the future and practise self-restraint for the benefit of some later reward had previously been touted as exclusive to humans and other primates. Moreover, the above-mentioned study indicates that chickens are, like humans, susceptible to anxiety and frustration.

THE SOCIAL INTELLIGENCE OF CHICKENS

Scientific studies of the interactions of chickens with each other, and with other birds and animals, have been conducted on wild jungle fowl in their natural habitats, feral domestic fowl in the wild and domesticated chickens in captivity or in the laboratory setting.[19] Despite the different circumstances of wild and domesticated fowl, their behaviours, when permitted natural expression, remain very similar. Selective breeding of domesticated chickens, in order to ensure maximum egg production or rapid growth for the chicken meat industry, has led to some behavioural changes, no doubt resulting from the physical peculiarities bred into these birds. Nevertheless, even broiler chicks and layer hens will struggle to satisfy fundamental gallinaceous desires within their restrictive environs (an issue revisited in chapter Six).

In the wild, chickens usually live in smallish groups comprising a dominant rooster and one or more hens. They prefer to flock together, adhering to clearly defined 'chicken etiquette' within their families and wider social formations. Their specialized visual, auditory and olfactory perceptions assist them in rapidly recognizing friends (members of the same group) and foes (foreign chickens, predators, threatening objects and situations).

There are seasonal variations in the membership of groups. In the breeding season, the dominant rooster of each flock marks out and defends his territory, either alone or accompanied by several non-broody hens. During this time other hens in the flock leave the group to lay eggs. The rooster helps locate and settle each broody hen in a nesting site. While she is incubating a nest of eggs, however, and for most of the time she is raising her chicks, a mother hen prefers isolation and maintains some distance from the flock. Once her offspring lose the down feathers from their faces, she initiates aggressive actions towards them in order to encourage them to move on from the maternally led group. When the brood is independent of her, the mother hen returns to the rooster and the original flock.

Each clutch of almost-grown chicks remains apart from others until, by a process called 'streaming', the various broods of a given season amalgamate into a single group prior to sexual maturity. Young roosters begin to leave this larger formation once they become sexually interested. They then encroach on the borders of other established roosters' territories, usually being careful to remain on the periphery. As the pullets reach sexual maturity they are increasingly chased and accosted by roosters until they obtain protection within the flock of a dominant rooster.

During non-breeding season, dominant roosters and their flocks inhabit overlapping home ranges with fixed hierarchies between neighbours. In each flock the older hens remain closest to the rooster, while younger ones lurk on the outskirts of the group. Subordinate roosters loiter on the periphery of flocks; the most inferior males are positioned furthest away, leading more or less solitary lives and remaining the most vulnerable to predation.

A hierarchical society – colloquially referred to as a 'pecking order' – is very important to chickens. In natural flocks the rooster is the protector of the hens and chicks and of the territory of the group. Individual hens also have a ranking within each flock. The pecking order is established more or less through trial

and error within same-sexed groups of chickens (it is unusual for hens to fight roosters, or vice versa). Young birds are subjected to aggression as they learn the 'propriety' of the flock, probably as a result of coming too close to older or more dominant birds. These same chicks may later become antagonistic towards other birds. Weaker members of the group are also regular candidates for hostility in the form of face-to-face fighting or head and neck pecking. As a rule, chickens learn quickly which birds they can beat and which will defeat them. Linear pecking orders are easy to spot, where a 'top' hen dominates all others, with the 'deputy' bird dominating those below her and so on; but many pecking orders are non-linear. Triangles may occur in which bird A pecks bird B who pecks C who pecks A; and square peck orders are also possible. Once a pecking order has been established it is maintained through recognition of individual birds and their relative positions.[20]

Crowing is an important feature of chicken courtship spectacles, along with behaviours such as waltzing, wing-flapping and titbitting. Waltzing describes the way a rooster circles a hen while keeping his outside wing lowered. Titbitting is a food-related display during which a rooster calls to a hen while he pecks at – or picks up and drops – items of food. Either demonstration may bring a hen closer to the rooster. As well as vocalizations and displays, certain characteristics of the head and neck convey information about identity and rank. For example, the size and colour of combs are influenced by levels of sex hormones, thereby imparting clues as to the status of dominant roosters. A hen's selection of a mate may be influenced by comb length and colour, eye colour and spur length,[21] wing-flapping displays,[22] or anti-predator alarm calling.[23] Roosters are also capable of clever manipulation and deception when it comes to the business of sex. Sometimes a hen is tricked into approaching a rooster who emits food calls even though no food is present; the rooster's subsequent behaviour suggests that he has drawn her nearer merely to increase the likelihood of sexual contact.

The prominent cognitive psychologist David Premack once remarked that 'even if chickens had a grammar, they would have nothing interesting to say'.[24] The more we know about chickens, however, the more premature this judgement seems.

Since the seventeenth century, when the French philosopher René Descartes' ideas about human exceptionalism became influential, we have assumed that language is unique to humans: our capacity to talk makes us special – and acoustic interactions in other species (with the exception of cetaceans, parrots and some non-human primates) have been attributed to elementary emotional states or involuntary or reflexive responses. Vocal communication between chickens has also been doubted because we may assume, like Premack, that if other creatures *were* capable of 'talking' to each other then what they are saying would surely be readily interpretable by humans. In contrast, Charles Darwin believed rudimentary forms of language could be detected in human language 'had evolved from precursors in the natural signals of animals'.[25]

In truth, chicken talk is a complex affair, involving visual, vocal, olfactory and tactile senses, combined to convey numerous intentions, messages and details amongst chickens. For example, roosters interact with each other to establish identity, status and proximity. Each rooster has a distinct crow, the acoustic frequency of which corresponds to the length of the bird's comb. Dominant roosters crow more frequently than subordinate males and may attack crowing roosters of lesser status. The ability to determine the superiority of an individual rooster by the quality or rate of his crow allows dominant males to identify other dominant adversaries, while subordinate roosters are able to maintain a 'respectable' distance in order to avoid attacks.[26]

The German 'fowl linguist' Erich Baeumer suggested in 1964 that domestic chickens spoke 'an international language made up of 30 basic sentences' or calls.[27] Over 60 years he recorded hours of 'chicken talk', later isolating individual 'sentences' so that these could be matched to his written accounts and to photographs of specific behaviours. Half a century on, avian specialists now accept that wild and domestic fowl produce at least 30 distinct forms of vocalization, including territorial, location, mating, laying and nesting, submission, distress, alarm and fear, food and contentment calls.

Chickens emit two types of alarm or warning calls, each produced in response to a specific type of danger. Aerial alarm calls comprise a series of low-intensity short, narrow-band whistles or screeches, and occur when a chicken identifies a predator, such as a raptor, approaching from overhead. They happen in the company of other chickens and often while the caller attempts to hide. Alarm calls in response to threats identified at ground level (cats, dogs, foxes and so on) consist of loud, repeated and conspicuous 'pulsed broad-band cackles', directed as much at the predator as at the other chickens.[28]

The presence and nature of an audience affects calling by chickens. Roosters elicit fewer alarm calls (or none at all) when alone or with unfamiliar hens than when accompanied by flock mates or young chicks. Similarly, roosters signal when food is discovered, particularly when hens are in the vicinity, and they call more when favoured foods such as worms and peas are present. Hens respond to both types of call but prefer calls inviting them to share the special treats.[29]

Chickens also demonstrate syntax and semantics, once thought to be the exclusive hallmarks of human language. In one experiment testing the complexity of acoustic signals in Sebright bantams, caged hens were played recordings made by roosters in response to the two types of predators, aerial and ground. No visual cues were given; the only information hens received came from the recorded calls. When hens heard aerial alarm calls they ran for cover, crouching and looking upward as they would in the natural environment when threatened by a hawk. When confronted by a ground alarm call, they took on 'unusually erect "alert" posture[s], becoming more active and scanning back and forth in a horizontal plane', as if searching for the presence of a dog or cat.[30] In a modified version of this experiment hens that were shown computer-generated images of possible predators emitted the strongest alarm calls when viewing large, fast-moving, bird-like shapes overhead, suggesting that chickens respond to precise characteristics of any threat. Alarm calls are therefore meaningful: rather than being involuntary exclamations simply reflecting a chicken's internal state of fear, they transmit quite specific information that will be understood by receivers. The exchanging of information about food is also precise; chickens respond to calls indicating the presence of novel food, but are not impressed by repeated food calls or messages about known food.[31] Thus it now appears that the cognitive processes involved in representational thinking in chickens are similar to those required for associative learning in humans.

THE EMOTIONAL LIVES OF CHICKENS

More often than not it is inferred that only humans – and perhaps certain creatures we grow fond of and close to – are blessed with the capacity *really* to feel. Chickens are quite easily dismissed as

The delight of dust-bathing.

impassive because relatively few people have experience of living alongside them; but those who do need no convincing of their individual responses, their likes and dislikes, and their distinctive personalities.

Chickens derive immense pleasure from dust-bathing. Wild fowl and free-ranging domestic chickens routinely engage in this activity, which involves finding a suitable patch of ground, scratching around until this site is prepared in the shape of a bowl, then hunkering down and using wings to scoop up dirt onto the feathers (a similar pastime is sunbathing, when chickens lie down and stretch out their wings towards the sun's rays). Chickens dust-bathe for half an hour or so, frequently adjusting position and rubbing their bodies further into the earth. When bathing is completed, the birds rise up and shake themselves vigorously. Dust-bathing functions practically to remove parasites and oil from feathers, but studies have shown that hens will work to repeat a dust-bath even when they have just had one, indicating that they seek this experience for its own sake.[32] Dust-bathing is also an important social event with flock members collecting together to engage in this enterprise, often splashing dirt across each other in a frenzy of bathing.

Indeed, just like other gregarious beings, hens form close relationships with particular members within their group, often foraging alongside each other and sharing titbits and nesting companionably when laying eggs. I have observed how, when one hen has been separated from her preferred flock mate (when laying, for instance), she will call out as she emerges from her nest, and both hens will then make their way towards each other and reunite independently from the rest of the flock. While this behaviour may be interpreted as one hen fulfilling the function of a surrogate 'escort' rooster in an all-female flock, it seems equally reasonable to understand it as a fondness and concern for the well-being of a particular friend.

If chickens experience close relationships with each other, it follows that they will respond to the absence of a special companion. In *Just Like an Animal*, zoologist Maurice Burton tells of the intense altruistic friendship of two hens, one old and almost blind, the other young and healthy. The younger hen protected them both, escorting her incapacitated companion around the garden during the day, collecting food for her and vocalizing its arrival, then leading her at nightfall to their roost. When the older hen died, her friend stopped eating and rapidly weakened. Within a week she too was dead.[33] The deaths in close succession of chickens who are devoted companions are a familiar phenomenon to those who live with and observe these birds.

While such close relationships between chickens seem natural enough, inter-species affection is also evident. In the late eighteenth century Gilbert White – natural historian, ornithologist and cleric (and considered a major inspiration for the environmental movement) – recorded his observations and beliefs about inter-species friendships in a letter dated 15 August 1775. Referring to a chicken and a horse, White wrote that

> these two incongruous animals spent much of their time together in a lonely orchard, where they saw no creature but each other. By degrees an apparent regard began to take place between these two sequestered individuals. The fowl would approach the quadruped with notes of complacency, rubbing herself gently against his legs; while the horse would look down with satisfaction, and move with the greatest caution and circumspection, lest he should trample on his diminutive companion. Thus, by mutual good offices, each seemed to console the vacant hours of the other.[34]

When it comes to friendships, chickens, like humans, are all different. Some prefer to be around chickenkind only, while others are inter-species extroverts, enjoying the company of humans and other animals as well as their own flock. In his sixteenth-century treatise on chickens, the ornithologist Ulisse Aldrovandi described the devoted friendship he developed with a particular hen: '[I] raised a hen who, in addition to the fact that she wandered the whole day alone through the house without the company of other hens, would not go to sleep at night anywhere except near me among my books'.[35] Five hundred years later a Swedish wildlife rehabilitator, Chatarina Krångh, shared a similar relationship for eight years with a Leghorn hen called Tikka who was rescued as a chick from an experimental hatchery. When Tikka developed an illness necessitating her removal from the other chickens living on Krångh's property, she was brought indoors for antibiotic treatment. Following her recovery, Tikka 'firmly refused' to return to the flock, instead establishing a home for herself in the house with her human companions. Krångh recalls:

> Tikka enjoyed going with me to work and when she was tired in the afternoon she went in to her cat box and waited for me to get ready to go home . . . She loved to lie by the soap-stone stove and when she went to bed early in the evening, she tried to get the rest of the family to join her.[36]

This story not only shows how very close relationships may be forged between humans and chickens (not to mention how adaptable birds can be), it also demonstrates how chickens are capable of asserting their own forms of agency and self-determination; in Tikka's case, this involved the enactment of

Bantams enjoying the sunshine.

a physical rejection of her own species in favour of life with Krångh and her family.

Just as chickens experience positive feelings such as the joy associated with dust-baths, sunshine and friendship, they also endure negative emotions including fear, anxiety, frustration and boredom. In nature, chickens make a sound called the gakel-call when they are frustrated; for example, when they are thwarted from obtaining desired objects, or from performing usual activities. (The gakel-call is also used by a broody hen to alert the rooster to her imminent departure in order to incubate eggs; the rooster then escorts her to the safety of a nest site.) In scientific studies depriving hens of food, water and dust-baths, chickens also elicit gakel-calls, and when confronted with additional obstacles, they emit more of these specific calls.[37] Chickens exposed to frightening conditions in the laboratory also register a phenomenon known as 'emotional fever' during which their core body temperature rises. The presence of emotional fever has been used to argue the case for consciousness and emotions in different species.[38]

And, of course, chickens suffer pain. There is some suggestion, however, that chickens (and birds in general) experience pain differently from mammals, attending to either physical discomfort or emotional distress, but not both at the same time. This notion is based on invasive experiments that tested pain and fear responses of arthritic chickens whose frontal brain lobes were removed. Injected with pain-causing substances, the chickens appeared to respond to pain only in the absence of any fear-producing distraction, leading the researchers to propose that 'chickens may suffer from chronic pain when they are undisturbed, but when disturbed or frightened, the pain ceases and the chicken can only attend to the fear.'[39]

As a trick at conferences, the avian behaviourist Chris Evans lists some of the perceptual, cognitive and communicative capacities of chickens discussed in this chapter, without revealing which species he is referring to. His audience invariably assumes he is talking about monkeys. Evans concludes that 'to the extent that our attitudes towards animals are shaped by their perceived mental life, such findings [about the abilities of chickens] should be thought provoking'.[40] As more is revealed about the distinctive wisdom of chickens, this may prompt not only a shift in our attitudes towards these birds, but also positive changes in our modern-day relationships with them.

It is also the case, however, that while chickens display feelings comparable to those of humans (such as grief, fear or happiness), they no doubt also possess their own exceptional forms of emotion and consciousness that even the most rigorous scientific tests may not begin to uncover – simply because these inimitable perspectives of chickens do not register conceptually or experientially within the human domain. Instead of fearing or dismissing this alterity, we might instead respect and take pleasure in the uniqueness of chickens, their inscrutable yet delightful chicken-ness, their complex and nature-loving chicken worlds.

The photography of artist Mary Britton Clouse focuses on human friendships with chickens.

3 Chickenlore

In some cultural traditions chickens have been respected and even revered; in others they have been feared or abused. Chickens have been worshipped as the representatives of deities, and persecuted as the conduits of evil. They have been sacrificed as offerings to the gods or as substitutes for human sin. The global repository of chicken and egg mythologies, proverbs and superstitions is vast: what follows is a survey of the prevalent themes – along with a glance at some less familiar beliefs.

SYMBOLISM OF THE EGG

One of the main philosophical questions posed since Aristotle's day relates to the cause and consequence puzzle of the chicken and the egg: which came first? In this chapter the egg comes first, because it has been prominent in creation narratives across the globe and is almost universally associated with new life and rejuvenation.

The mythical cosmic or world egg is an ancient example of the symbolic linking of eggs to birth. Generally, the cosmic egg embodies 'the idea of a silent universe, all at once bursting into activity and chaos'.[1] In some creation narratives the cosmic egg incubates the universe and its inhabitants: its yolk forms the

A Greek terracotta warrior astride a hippalectryon (half horse, half rooster); 500–470 BC.

ground and its white the sky (or vice versa); in other accounts the egg emerges from the primeval waters or it is a 'sun egg' hatched by the warmth of the sun.

The first recorded reference to a cosmic egg occurs in an Egyptian papyrus of the New Kingdom, possibly connected with Thoth (God of the moon), who appears as an ibis and either hatches the world egg at Hermopolis or emerges from the egg himself. In the Chinese myth of P'an Ku, dating from around 600 BC, it is said that 'the space of the universe was in the shape of a hen's egg'.[2] Within this cosmic egg (representing 'the principle of fertility, wholeness and duality') there was *no thing*, and inside *no thing* the first being, P'an Ku, developed and eventually burst forth, splitting the eggshell in two. The lighter half (yang, the egg white) rose to form the sky, while the heavier half (yin, the yolk) fell to create the earth.[3] Similarly, the creation myth of the Japanese *Nihongi* (a Shinto document dating back to 8 BC) imagines an initial chaotic mass like an egg within which the properties of masculine and feminine were indistinguishable.[4] The Indian *Rig Veda* conveys the story of the mighty creator Prajapati who fertilized the primeval waters, producing a Golden Egg from which, after one thousand years of gestation, Brahma emerged (along with the continents, oceans, mountains, planets, gods, demons and humanity).[5]

In many of these narratives the egg acts as an incubator, inside of which gestates the Creator or equivalent, the source of all life. A Yoruba creation myth of *c.* 300 BC, however, focuses on the life-bringing activities of the hen herself, whose scratching brings forth the world. A white hen, freed by the young god Obatala, lands on sandy waters, 'immediately beginning to scatter the sand by scratching at it . . . Wherever the sand fell, it formed dried land. The larger piles of sand became hills, while the smaller piles became valleys.'[6]

Sonja Rooney,
*Yoruban Creation
Myth*, 2009.

In Europe, until the Middle Ages, the earth itself could be understood as an egg incubating precious metals. An alchemist wanting to create the philosopher's stone, a substance capable of transforming common metals into gold or silver, would use an egg-shaped crucible called the philosopher's egg because of the symbolic associations with the four elements essential to alchemy: earth (shell), water (egg white), fire (yolk) and air (membrane).

Traditionally, the egg has been honoured during ceremonies welcoming spring and the arrival of new life after cold, dark winters. Today's consumers expect eggs to be available on

Alle gute ding seÿnd dreÿ
Drum schenck dir dreÿ Oster · Eÿ ·
Glaub und Hoffnung samt der Lieb
Eÿentahls auß dem Herden schirff
Glaub der Kirch, vertrau auf Gott
Liebe Ihn biß in den todt.

C.P.S.C.M.   G. B. Göz inv. et excud. A.V

demand, but until relatively recently eggs were plentiful only for a few months of each year. The spring equinox marked the reappearance of fresh eggs – and broody hens. In ancient times, springtime was associated with the Germanic goddess of fertility, Ostara (or Eoestre). Accompanied by her consort, a human-sized or regular-sized rabbit, Ostara delivered fresh eggs in the Northern springtime, giving her name to the festival of Easter, which is still associated with rabbits and eggs. On the instruction of Pope

Gregory the Great, the Church colonized pagan folk customs, including Easter: the egg then came to represent the resurrection of Christ, and of Christians, from the dead.[7]

The centuries-old English spring ritual of pace-egging (from the word *Pasch*, Old English for Easter) involved rolling down hillsides eggs that had been decorated with onion skins and boiled to a golden colour. On Easter Sunday eggs were given to the street performers who sung pace-egging songs before enacting 'mumming plays' (seasonal folk dramas), a tradition persisting to this day in Lancashire and Yorkshire. Of European medieval origin, games such as egg dancing and egg tapping were also performed; the former required dancers to step around clusters of eggs without breaking them, the latter involved holding onto one's own

'Egg Rolling on the White House South Lawn', Easter Monday 1929.

egg securely while trying to crack open a competitor's.[8] To this day, Easter Monday is marked in the United States via the annual Egg Roll held on the White House lawn. This event involves children pushing an egg through the grass with a long-handled club.

Eggs have been coloured, to symbolize life, for thousands of years and in myriad cultures. Across Europe red is the most prominent dye for Easter eggs; Romanians say 'red eggs at Easter' to describe something inevitable, because in their country it would be unthinkable to have Easter without these symbols of the Resurrection. In China, red eggs are given to children on their birthdays to foster long life and happiness. A different custom from Ukraine involves eating eggs called *krashanka*, which are dyed bright yellow to represent the rebirth of the sun.

Alongside the ubiquitous association of eggs with new life sits an almost opposite symbolism: the egg as an emblem of fragility and breakability. The most famous personification of this connection is Humpty Dumpty – known as Boule Boule in France and Lille Trille in Sweden and Norway – whose fate is described in the popular rhyme first printed in 1810:

Humpty Dumpty sat on a wall.
Humpty Dumpty had a great fall.
All the king's horses and all the king's men
Couldn't put Humpty together again.

There are many hypotheses about the origins of this rhyme, but the relationship between 'Humpty Dumpty' (eighteenth-century slang for a short, dumpy, clumsy man) and eggs was permanently cemented by Lewis Carroll's *Through the Looking-Glass* (1871), in which Alice comes across Humpty Dumpty on a wall and offends him by mistaking him for an egg. Pompously, he explains that he has no fear of falling because the sovereign

has promised to use all his horses and men to pick him up again. As Alice leaves him she hears 'a heavy crash [that] shook the forest from end to end'.[9]

So it is that the egg concisely encapsulates the most profound and potent of paired meanings: the simultaneously opposed and linked ideas of creation and destruction, or death and resurrection. Even the mundane axiom that one 'can't make an omelette without breaking eggs' draws on these connotations since it signifies the necessity of breaking out of an old state to attain a new one. Similarly, in cultures throughout the world and throughout history, the egg's two salient meanings of destruction and creation have made it a potent source of folk belief and medicine.

Ukraine is the home of *pysanky*, the ancient practice of elaborately decorating eggshells with beeswax and bright dyes, and giving these as gifts to those one admires or loves. It is said that 'as long as women create pysanky, the powers of life prevail [but] when the last woman to make pysanky stops doing so, then evil will reign triumphant over Earth.'[10]

An old British rural custom declares the first egg laid by a pullet to be lucky. If the hen is brown a wish should be made when the egg is eaten; if the hen is white the egg should be placed under a pillow and the sleeper will be granted a vision of his or her future lover. A black pullet's first egg guarantees safety from fever for a year. Collecting eggs after sunset is bad luck.[11] So is finding a miniature egg, because it may have been laid by a rooster and therefore contain a gestating serpent called a Basilisk, which can be prevented from hatching only by throwing the offending egg over the barn roof.

For centuries people have debated which end of an egg is best to crack open: contemporary folklore considers it ill luck to crack open the smaller end. In Jonathan Swift's *Gulliver's Travels* (1726) violence erupts between the empires of Lilliput and Blefuscu over just this dilemma (a satire on the political and religious squabbles of Swift's time).[12] If, having cracked your egg, you discover it contains two yolks, you can expect an upcoming wedding – that is, if you are in Somerset; in the north of England double-yolkers portend bad luck.[13]

In some South-East Asian and Pacific Island cultures, preference has been shown for consumption of brooded eggs containing well-developed embryos. The belief that such eggs are delicacies may arise from the egg's symbolic link to fertility and a fear that it is dangerous to eat eggs before they contain a recognizable life form (in contrast, rejection of eggs as food in parts of Africa relates to the notion that they are excrement from hens).[14]

The shells of eggs are also riddled with potential mischief. Disposing of eggs should be done carefully: never burn eggshells lest your hens cease laying. A Japanese superstition forbids women from stepping over eggshells for the sake of their sanity. A belief prevailing since the time of Pliny the Elder (AD 77) cautions people to shatter the shell of a consumed egg in order to prevent its use in sympathetic magic against the person who has eaten its contents; while a similar custom, reported across the United Kingdom since at least the 1580s, requires anyone having eaten a hard-boiled egg immediately to shatter the shell. This is because resourceful witches use discarded eggshells as boats from which they head to sea and brew up storms for sailors. Similar fears may be behind an old fishing custom forbidding eggs, and even the uttering of the word 'egg', aboard boats.

There are many spells, charms and curses involving eggs. These may be cast to bring about fertility, abundance and prosperity, as well as healing, protection and purification. Sympathetic healing magic entails moving illness or misfortune from a human being to another living being or an object. A traditional

Century eggs are a Chinese delicacy made by preserving chicken eggs in a mixture of clay, ash, salt, lime and rice hulls for several months. The yolks turn into a dark green substance smelling of ammonia and sulphur while the white becomes jelly-like.

French cure for a sick child involved feeding a dog an egg filled with the child's urine, while in Germany jaundice could be cured by giving a hen an egg containing the affected person's blood.[15] The Ait Wairain of Morocco believe that sleeping under the stars can lead to white spots in the eyes: the cure is to inscribe words from the Koran onto an egg, which is then touched to the eye before being cracked open and emptied. If a white spot is evident in the contents of the egg the person will get better. In modern-day America some believe that rubbing eggs across birthmarks each morning will cause them to disappear, but only if all the eggs are then buried under the doorstep.

An ancient Muslim remedy for the evil eye involves waving egg, salt and turmeric over the affected person, then throwing these ingredients down at a crossroads. In New Mexico the name *El ojo* ('the eye') is given to a fever caused in children when they are shown an excess of affection. The suffering child is confined to bed-rest with the contents of an egg placed on a plate near the head. *El ojo* will heal once an eye has appeared on the egg. Across the world eggs have also been used to drive away demons: in some parts of China, for example, devils are asked to enter eggs during exorcisms.[16]

Eggs are employed in fortune-telling too. Oomancy involves predicting the future by reading the inner and outer forms of the egg, for example, by breaking an egg into water and interpreting the shapes formed by the white. Oomantia refers to the reading of signs appearing in eggs, such as the clotting patterns formed after an egg white is left exposed for a day.

ROOSTERS AND MASCULINITY

Human assumptions about masculinity and femininity dominate cultural depictions of roosters and hens. Since ancient times, and across many cultures, roosters have customarily been portrayed as strong, brave and vigilant leaders, mates and fathers – steadfast in their protection of a flock and its territory. (The Jewish Talmud encourages people to learn 'courtesy towards one's mate' from a cock, presumably since roosters call hens to food and wait for them to eat first.) Hens have been represented as the epitome of maternal devotion and domesticity, willing to sacrifice themselves for the safety and well-being of their brood. When chickens transgress these gendered assumptions – for example, when roosters sit on eggs or hens crow – they give rise to superstitions and urban legends. In Switzerland in 1474 a rooster found guilty of heresy after allegedly laying an egg was burned at the stake along with his egg;[17] while traditional folklore from Germany to Persia stipulates that any hen who crows

Harry Kerr, 'Indian Rooster' in the *Ornithologiae* (1599) of the Italian naturalist Ulisse Aldrovandi.

must be killed immediately because she portends disaster.[18] (In fact, it is not unusual for hens to crow, especially if they live in a flock without a rooster.)

In those areas of the world where chickens were first domesticated, such as South-East Asia, India, China and Persia, the original value and use of roosters probably related to their strident crowing, which acted as a natural alarm clock to signal the start of day, and also as a natural alarm system, warning of

THE COCK AND PRECIOUS STONE.

F.Barlow delin.   Ja.Kirk fcit.

A Cock in fearch of Food the Dunghill tries,
A fparkling Jewel gliftens in his Eyes;

Cry'd he _ A Barley-corn wou'd pleafe me more
Than all the Treafures of the Eaftern Shore.

MORAL.

Gay Nonfenfe does the noify Fopling pleafe,
Beyond the nobleft Arts and Sciences.

Francis Barlow's depiction of Aesop's fable of *The Cock and the Precious Stone* (c. 1670). 'A Cock, scratching for food for himself and his hens, found a precious stone and exclaimed: "If your owner had found thee, and not I, he would have taken thee up, and have set thee in thy first estate; but I have found thee for no purpose. I would rather have one barleycorn than all the jewels in the world".' (interpreted by George Townsend, 1867).

disturbances or intrusions. Roosters were also kept for their assumed magical properties: in ancient Egypt this meant that they were sacrificed to deities; in India *c.* 1000 BC their special religious status gave chickens immunity from being eaten, although parts of their bodies may still have been harvested for their mystical qualities.[19]

The practice of cockfighting, presumed to be around 3,500 years old, arose in some regions because of the rooster's role in divination. It is also speculated that cockfighting as a form of human competition occurred, rather ironically, because people observing natural cockfights deemed these to be unfair confrontations, with one rooster clearly superior in physique and energy. Originating in South-East Asia, cockfighting spread – as chickens did – to China, India, Iran, Greece, Rome, western

Caged roosters at a market in the Philippines.

Europe (including Britain), the Caribbean and the Americas. It remains immensely popular in parts of Asia, Europe, America and the Caribbean and is the national pastime for men in the Philippines, Bali and Puerto Rico.

Under natural conditions roosters fight each other to establish dominance over territory or flocks, but these aggressive encounters rarely result in death or even serious injury because they cease once a victor is apparent. In contrast, artificial cockfighting involves two purpose-bred roosters set against each other within a confined space (typically a small circular pit) and usually forced to fight to the death. Each bird is fitted with steel spurs called gaffs that have round blades ranging in length

Pierre Brunet-Houard (1829–1922), *Fighting Cocks and a Hen*, oil on canvas.

from one to three inches (2.5–7.5 cm), which curve to very sharp points. Sometimes 'slashers' are used instead of gaffs: these instruments have sharp edges the entire length of the blade. In the Philippines knives or *tari* are tied to a rooster's left foot. In addition, the combs of fighting roosters may be cut back so status as conspecific or predator is confusing to opponents. Considered a sport by the men involved, cockfighting is viewed as atrocious cruelty by those concerned with the birds' welfare. Even in those countries where cockfighting has been banned on humane grounds, however, it persists as an illegal underground practice.

Cockfighting has inspired poetry, literature and art. Popular sayings derive from cockfighting slang such as 'to turn tail', 'to raise one's hackles' and 'to show the white feather'; and certain metaphors – 'to be cocky' or 'cocksure' – allude to the association of cockerels with 'cocks' or penises. Indeed, cockfighting is an overwhelmingly masculinist activity. Acting as a surrogate for its human trainer, a rooster's death is experienced as personal failure by a handler (or 'cocker'), while a victory confirms his manhood.

The folklorist Alan Dundes conducted one of the first psycho-analytically influenced cross-cultural examinations of 'gallus as phallus'. He believes several symbolic themes or events endorse the interpretation of cockfighting as male phallic combat: for example, the phallic association of cocks (the birds) with 'cocks' (penises) can be identified in the presence of cockerels placed 'atop penile Christian architectural constructions such as church towers'. Moreover, 'resurrection, if understood as re-erection, or even in the narrow Christian sense of rising miraculously from the dead, is perfectly understandable in cockfight terms'.[20] Roosters injured during fighting may be 'revived' by their owners in order to continue the competition. In the Philippines this involves inserting chilli into a weakened rooster's anus, while in Bali red pepper is stuffed down a cock's beak or up his anus 'to give him spirit'. Alternatively, cockers blow air into the mouths or the anuses of roosters. Among the Tulu, the act of blowing in a rooster's anus has even become a humorous metaphor of everyday life.[21]

Such activities prompt Dundes to read cockfighting as 'a sublimated form of public masturbation': in addition to the act of blowing and sucking on the cock's head (as a means of revival), cockers spend considerable time stroking, cleaning, bouncing and rubbing their roosters, as well as massaging the necks of the birds. As further evidence for this interpretation, the American slang saying 'to choke the chicken' is a standard euphemism for masturbation.[22] The blending of sexuality, gender and violence means that homoeroticism, homophobia, masculinism and misogyny are all simultaneously at work in the cockfight. The fight itself is a contest of masculinity: the winning bird (and, by association, his human owner) is endowed with virility and strength, while the losing bird suffers penetration by the spurs of the other (a displaced form of emasculation or feminization for his owner). In Malay two roosters destined to fight each other are said to be 'betrothed' and the fight involves testing which bird will become 'the female' of the pair through weakness

Cockfighting match in the Babur Gardens, Kabul, 2008.

and failure. The phrase 'Vamos, como tu padre!' ('Let's go, like your father!') can be heard at fights in Venezuela when urging one cock on to win, while in Brazil onlookers chastise the losing bird by crying 'the mother's blood is showing!'[23] In Mexico, the term *gallo-gallina* (rooster-hen) is used to accuse one of cowardice or of homosexuality.

THE HEN AS IDEAL MOTHER

What of the hens whom we observe each day at home, with what care and assiduity they govern and guard their chicks?
Roman historian Plutarch, AD 46–120[24]

Thanks to human prejudices about motherhood and domesticity, the hen has been afforded a little varied and low-status role in mythology and folklore. Nevertheless, for many centuries hens have been respected for their maternal devotion and nurturing qualities, as well as for the eggs they produce. In his comprehensive treatise on all matters pertaining to fowls and their eggs, the Italian ornithologist and chicken enthusiast Ulisse Aldrovandi (1522–1605) expressed his admiration for the loyalty, vigilance and affection of the hen towards her family:

They follow their chicks with such great love that, if they see or spy at a distance any harmful animal, such as a kite or a weasel or someone even larger, stalking their little ones, the hens first gather them under the shadow of their wings, and with this covering they put up such a very fierce defense – striking fear into their opponent in the midst of a frightful clamor, using both wings and beak – that they would rather die for their chicks than seek safety in flight, leaving them to the enemy. Thus they present a noble example in love of their offspring, as also when they feed them, offering the food they have collected and neglecting their own hunger.[25]

Seen as an embodiment of maternal power, the hen with her eggs and chicks has sometimes even been regarded with awe, as demonstrated by the creation myths discussed at the start of this chapter. Ornery hens, on the other hand, have triggered concerns about wayward women: among the Sema, a tribe inhabiting Nagaland in north-east India, women are not permitted to eat hens that 'lay here and there in different places' for fear they will become rebellious and unfaithful; while the Nguni people of South Africa believe it is the consumption of hens' eggs – viewed as aphrodisiacs – that may entice women to behave lasciviously.[26]

RELIGIOUS AND SPIRITUAL SYMBOLISM

There are other ways, too, in which chickens feature, sometimes very centrally, in a variety of religions, both ancient and modern. For instance, the rooster (along with the snake and pig) appears in the middle of the Buddhist Wheel of Life.[27] Many belief systems associate chickens, particularly roosters, with the sun (lightness and good) or the moon (sometimes associated with darkness and evil). Indeed, it has been argued that cocks embody 'three of the most powerful themes in nature – sexuality, the sun, and the theme of resurrection'.[28]

The fourteenth-century BC pharaoh Akhenaten mentioned the chicken in his Hymn to the Sun. The Talmud proclaims: 'Praised be Thou, O God, Lord of the world, that gavest understanding to the cock to distinguish between night and day',[29] while in the Hadith Muhammad announces: 'When you hear

a cock crow, ask for Allah's blessings for he has seen an angel.'
During the Kianian period in ancient Iran (2000–700 BC) the
rooster was the most revered of birds: sacred to the Persian sun
god Mithras, his crowing heralded the dawn and dispelled the
evil spirits of night-time.[30] The Gnostic being Abraxas is a powerful sun-related entity with the head of a rooster and serpents
for legs, who symbolizes good (God) and evil (the devil), as well
as other dualisms, in the one being.[31]

In East Asia the cock occupies the tenth position of the Chinese zodiac and symbolizes yin (feminine energy, night time,
coolness). Legend has it that Buddha asked all the animals to a

A religious sculpture in the Saddar cave near Hpa'an, Myanmar.

meeting on New Year's Day, rewarding the twelve who showed up with a 'time' – as well as a month and a year – named in their honour. Cockerels were given the 'You hour' from 5 to 7 in the evening, the time when they were thought to be busiest (because they were returning to the roost). Humans born in the year of the Rooster are said to be eccentric, theatrical, humorous characters, possessing good organizational and financial skills and enjoying being the centre of attention. The Day of the Chicken is the first day of the lunar Chinese New Year, so called because only the Heavenly Chicken knows when the first sunrise of a new year occurs. Traditionally at this time people would install chicken talismans on the doors of houses – to welcome the sun and the imminent springtime, and also because such talismans warded off demons. This custom continues in some rural areas of China today.[32]

Jesus mentions both roosters and hens in the New Testament. Foretelling his betrayal by the apostle Peter he declares: 'I tell thee, Peter, the cock shall crow this day, before that thou shalt thrice deny that thou knowest me' (Luke 22:34). Elsewhere, Jesus compares himself to a mother hen: 'O Jerusalem, Jerusalem, thou that killest the prophets, and stonest them which are sent unto thee,

Chinese New Year celebrations for the 'Year of the Chicken', 2005.

CANTIQUES SPIRITUELS.

Air: Ordinez-nous, Marie.

Du séjour de la gloire,
bienheureux, dites-nous,
après votre victoire,
quels biens possédez-vous?
Coelui songeant l'ineffable!
le cœur n'a point compris
quels trésors admirables
Dieu garde à ses amis.

Mais daignez-nous instruire
du prix de vos vertus;
dites ce qu'on peut dire
du bonheur des élus.
Loin du trouble et des larmes,
voir, aimer le Seigneur,
en jouir sans alarmes,
c'est-là notre bonheur.

Martyrs, dont le courage
triomphez des bourreaux,
quel est votre partage
après de si grand maux?
Tous la couronne reste,
la palme dans les mains,
sa terreur de délices
par quels nouveaux miracles

Dieu frappe-t-il vos yeux?
Ah! quel bonheur extrème
d'aller m'adresser
dans le sein de Dieu même
puiser la vérité!

Vous, humbles solitaires
que l'Égypte a produits,
de vos jeûnes austères
quels sont enfin les fruits?
Pour tous nos sacrifices
et nos saintes rigueurs,
sa torrent de délices
vient inonder nos cœurs.

Vous, épouses fidèles
du plus fidèle époux;
pour des ardeurs si belles,
quels plaisirs goûtez-vous?
Épouses fortunées,
nous pouvons en tout lieu,
de roses couronnées,
voir l'agneau de Dieu.

Vous qui de riche avare,
éprouvez les froideurs;
compagnons de Lazare,
qu'elles sont vos douceurs?
Nous mangeons à la table
du Roi de l'Univers;
le riche impitoyable
est au fond des enfers.

Et vous, qu'un pas de larmes,
nourrissant chaque jour,
quels sont pour vous les charmes
du céleste séjour?
Une main secourable
daigne essuyer nos pleurs,
un repos désirable
succède à nos douleurs.

Mais quelle est la durée
d'un si charmant repos?
Dieu l'a-t-il mesurée
sur celle de nos maux?
Dieu, qui de nos souffrances
abrège les douceurs,
veut que nos récompenses
durent dans tous les tems.

Ah! daignez-nous apprendre,
en cet exil cruel;
quelle route il faut prendre
pour arriver au ciel.
Si vous voulez nous suivre,
marchez en combattant
et sans cesse de vivre
mourez à chaque instant;

Mais la peine est extrème,
comment vivre toujours
en guerre avec soi-même
et mourir tous les jours?
Si la route est fâcheuse,
le trône est plein d'appas;
une couronne heureuse
pour de légers combats.

Air: De l'Hymne Confessor.

Ce saint confesseur de la
divine loi,
dont nous célébrons, et la
zèle et la foi,
en ce jour montz, triomphant et joyeux,
au plus haut des cieux,
Prêt qu'il a remporté,

Dieu toujours ardent,
sobre, chaste, pur, pieux,
humble, prudent;
rien n'a pu troubler dans sa
rare douceur,
la paix de son cœur.

Près de son tombeau, de
pauvres languoreux,
venez l'invoquer dans leurs
besoins pressans.
Dieu, pour décider quelle
est sa sainteté,
leur rend la santé.

Unissons nous tous et de
voix et de cœur,
chantons hautement
cet hymne à son honneur,
afin qu'à son besoin
nous ressentions toujours
son puissant secours.

Saint, gloire, honneur
prodant l'éternité,
au Dieu souverain, un en
sa Trinité
qui du haut des cieux, par
ses ordres divers,
règle l'univers.
Amen.

A Évreux, de l'Imprimerie
d'Ancelle fils.

ORAISON A SAINT-PIERRE.

Seigneur, sanctifiez et gardez votre Peuple, et faites qu'étant aidé par l'assistance de ce grand Saint, il vous soit agréable par le réglement de sa vie, et qu'il vous serve dans la tranquillité d'une sainte confiance. Amen.

A CHARTRES, chez GARNIER-ALLABRE, Fabricant d'Images; Libraire et Papetier, Place des Halles, N.º 17.

how often would I have gathered thy children together, even as a hen gathereth her chickens under her wings, and ye would not?' (Matthew 23:37).

A 19th-century print of St Peter the Apostle with a rooster.

Early Christians associated the morning crowing of roosters with Christ's vanquishing of darkness, sin and death. Up until the nineteenth century it was common for cockerels to be buried in the foundations of a church to keep evil at bay. As symbols of the Resurrection they were often carved on church steeples, and shaped into weather vanes, where they could turn their famed vigilance in all directions.[33]

While chickens are more commonly associated with the sun, they also are linked to the moon and to darkness. In ancient Greece black roosters were sacred to Hecate, goddess of the night, the spirit world and witchcraft,[34] while in Egypt they were sacrificed to Osiris, god of the afterlife. Pythagoras warned against killing white cocks because of their connection to the moon (on Easter Island white roosters were considered powerful conduits of sorcery, while Zoroastrianism declared them to be protectors against dark forces).

For centuries chickens have died for our religious or spiritual beliefs. In the orthodox Hasidic custom of Kapparot (kaparos), which precedes Yom Kippur, followers use kosher chickens as symbolic objects for the transference of their sins. In the 'Chicken Swinging Ritual' roosters are ritually sacrificed by men and hens by women. The observer swings the chicken around his or her head while reciting incantations symbolizing the passing of human sins to the bird. The chicken is then killed and his or her carcass may be delivered to poorer communities for consumption.[35] While many contemporary Jewish communities swing money for charity instead of birds, thousands of chickens are still slaughtered in New York each year among the more traditional followers of this custom.

Preparations
for the Jewish
Yom Kippur,
Day of Atonement.

Chickens are also sacrificed in Santeria, a form of spiritu-
ality established by Africans forcibly moved to Cuba during the
slavery trade. The dead chickens are offered as food to the *orishas*,
Yoruban spiritual entities now associated with Christian saints.
Chickens are also killed at ceremonies for the *loa* or gods in
Vodou religion (an African-derived Haitian tradition) and during hoodoo (magic). In India, although some Hindu laws forbid
consumption of their flesh, a chicken is slaughtered (along with
a duck, lamb, goat and buffalo) during the Panchabali or 'five
animal sacrifice' ceremony to appease Kali, the Hindu goddess
of death and destruction.

A NIgerian wooden figurine showing the attributes of wealth (the horse), fertility (the woman and child) and power (the many followers). An *orisha* (or Yoruba deity) might bestow these gifts on a devotee in return for the sacrifice of a cock.

Bahuchara Mata, Hindu goddess and patroness of the *hijra* (transgender and intersex) community in India. Bahuchara Mata carries a sword and flower in her two right hands and the scriptures and a trident in her two left hands, and sits on a rooster, this bird symbolizing innocence.

In Indonesia, Hindu cremations do not involve chicken slaughter, although a chicken may be present to act as a non-human repository for any evil spirits attending the ceremony. Following cremations, chickens are returned home to resume their usual lives.[36]

MAGIC AND DIVINATION

Chickens play a role in divination for many cultures. The Azande tribe of Sudan feeds chickens benge, a substance poisonous to them, and determines the answer to a problem depending on whether or not the birds survive. Africans taken to America as slaves brought with them beliefs about divination and magic; one tradition deemed that 'the avid scratchings of a wildly feathered, "frizzled" chicken in the earth [revealed] that some dangerous, unknown maleficence lay beneath the surface'.[37] The Pwa Ka Nyaw Po of Myanmar and Kariang or Yang tribes of Thailand use bamboo splinters inserted in the holes of chicken bones during their fortune-telling ceremonies, and decisions are based on the angles that the splinters arrange into when placed inside the perforations.[38]

The role of chickens as soothsayers goes back to ancient times. In Greek legend Alectryon was sent to guard against intrusion while Ares and Aphrodite conducted an illicit affair. Because he fell asleep at his post, leaving the two lovers to be discovered by Helios the sun, Alectryon was turned into a rooster so that he would forever remember to announce the coming dawn. Fortune-telling involving chickens is therefore called alectromancy, from the Greek *alectruon* for 'cock' and *manteia* meaning 'divination'. Alectromancers interpret chicken entrails during a ritual named *haruspicy*; they examine stones found in the stomachs of roosters (a practice called *alectorii*); and they observe chickens' appetites through a process called *oraculum ex tripudio*. For the latter method a circle is created using the letters of the alphabet, each sprinkled with an equal quantity of grain. A rooster or hen is placed in the circle and observed to see how quickly he or she eats and which letters are favoured. The letters chosen spell out a prophecy – a gallinaceous precursor of the Ouija board.[39]

The Etruscan civilization of ancient Italy and Corsica used another version of alectromancy: a high priest interpreted the order in which hens pecked at corn. The Etruscans also originated the wishbone custom – the practice of pulling apart the

chicken's dried clavicle bone to determine who gets the longer fragment and is thereby granted a wish. Both traditions were adopted by the Romans when they colonized Etruscan society at the end of the sixth century BC.

The armies and navies of ancient Rome travelled with chickens, not only for the eggs but also to foretell battle outcomes. During the Punic Wars, just before the sea battle of Drepana (249 BC), sacred chickens on board a Roman flagship were offered food to determine the best course of action, but displayed no appetite for the grain given them (possibly due to seasickness). Enraged, the consul P. Claudius Pulcher threw the birds overboard, declaring: 'If they won't eat, let them drink!'[40] The Romans were defeated – and the chickens vindicated.

In those regions of the globe where they have played an especially prominent role in human life, chickens are sometimes immortalized by their appearance in supernatural forms. Such mystical entities replicate several of the chicken motifs already discussed in this chapter, such as the association of serpents or snakes with chickens, the dualistic nature of chicken folk entities (both good and evil manifesting in different versions of the same creature), the prominence of roosters as warriors or protectors, and the transgression of 'gender norms' by chickens as ominous.

The ancient cosmography *Guideways through Mountains and Seas*, compiled between the fourth and the first century

Porcelain ewer with dragon spout and handle, a cockerel on the lid and chickens on the glaze, Mikawachi, c. 1890.

BC, showcases mythical creatures thought to have inhabited the Chinese countryside, oceans and rivers. Several of these beings are chicken-like in appearance or behaviour, such as the elusive Fuxi-Bird (a rooster with a human face) who resides on Baihao mountain, makes a sound like his name, and if seen by people, signals the start of war; the human-eating Qique or Qi-Magpie of North Shouting Mountain who sports several white chicken heads, the feet of rats and the claws of tigers; and the river-dwelling Shuyu or Shu-Fish, actually a red-feathered chicken with four heads, six feet and three tails, whose flesh cures melancholy when consumed.[41]

An even older mythical Chinese chicken is the Fenghuang, also known as the August Rooster because it may substitute for the Rooster in the Chinese zodiac, and is sometimes mistranslated as the Chinese Phoenix. The Fenghuang (whose Japanese version is called Ho'o) dates back to at least the Shang dynasty (1600–1045 BC), and is noted in later Zhou texts as 'a primary omen of political harmony'. It is depicted as 'five-colored and resembling a chicken with the markings of the graph for *virtue* on its head, *duty* on its wings, *ritual* on its back, *humaneness* on

its breast, and *trust* on its stomach'.[42] Because this entity combines both feminine (lunar yin) and masculine (solar yang)
qualities, it is thought to be positive, offering peace and beauty.

Europe also boasts a vast repertoire of supernatural chicken
folk. The Slavic fairy tale of Vasilissa the Beautiful features the
ambiguous figure of Baba Yaga, sometimes sought after for
her grandmotherly wisdom, but more often avoided on pain of
death, her darkest incarnation being a bogey-woman who eats
children (for example, she wishes to devour Vasilissa). Baba Yaga
lives deep in the forest in a cabin without windows and doors
that stands on chicken legs. While the witch on her mortar can
enter the house down the chimney, others wishing access must
chant 'Hut, o hut, turn your back to the forest, your front to me'.
At this command, the house rises on its legs and moves into a
position which allows the visitor to see a door. It is speculated
that the hut's chicken feet are associated with certain Slavic healing rites involving hens and chicks; in Russian tradition, for
instance, a pregnant woman may be warned off collecting eggs
from henhouses lest her baby develops 'hen illness' (constant
crying); and children affected by this illness might be taken to

Sarah Forgan, *Kikimora* (the Slavic chicken-like house spirit).

the chicken coop, which is asked to restore sleep to the child by giving her or him 'a human life' instead of a 'chicken life'.[43]

One of the most mischievous chicken entities of Eastern Europe is the scrupulous house-spirit Kikimora, who takes the form of a peasant woman with a long beak and chicken's legs and feet. If she is disappointed with the cleanliness of a home Kikimora prevents its occupants from sleeping by whistling in their ears or tickling their feet, but if she is pleased with what she finds she

Ivan Bilibin's depiction of the folklore figure 'Vasilissa the Beautiful' in front of Baba Yaga's house on chicken legs.

spends her time outside among the chickens.[44] In Slavic societies chickens are also fortunate to have a protector, Kirinyi Bog (the chicken god), who manifests as a hag-stone – a stone in which a hole has naturally appeared – and protects chickens from spirits such as Kikimora (and her moody husband, Domovai).[45]

The Liderc Nadaly, also known as 'the miracle chicken' (or *csodacsirke*), is a Hungarian vampiric creature who hatches from the first egg of a black hen placed under the armpit. Whether the creature takes chicken or human form, one of its legs is always that of a goose. Like many Eastern European folk figures, the Liderc Nadaly has both positive and negative aspects: in its demonic form it is an incubus for men and a succubus for women (literally loving its victims to death); in its benign form this creature tirelessly helps its human companion find gold or other treasure.[46]

Mystical rooster figures manifest in heroic and villainous forms across cultures. In French folklore, the Demi Coq (translated as Half-Chick) is the half-grown runt of a batch of hatched eggs. Considered a champion of the underdog because his 'cunning is overlooked by opportunists who only see his puny size', Demi Coq exists in Spain under the name Mediopollo and in New Mexico (where he may be a legacy from Spanish colonists) as Half-Rooster.[47] In Norse mythology Vithafnir is the great golden cockerel who perched atop the Scandanavian world tree (Yggdrasil) to guard it from enemies, and whose crowing portends the apocalypse on the day of Ragnarok.[48]

In Colombia, however, Pollo Maligno (literally translated as 'evil chicken') is the harbinger of misfortune. Said to appear whenever two or more horse-riders congregate, cheeping and following them relentlessly, people fear hearing the tweeting of Pollo Maligno because, depending on how serious he sounds, his chirps may foretell the death of a child. Those out walking may

also fall victim to Pollo Maligno since he will tweet incessantly until they go mad and return home. Only praying the novena of Archangel Michael will dispatch this malevolent chicken.[49]

CHICKEN MEETS SERPENT

A recurrent manifestation of mythical chickenkind – from the Basilisk of ancient Europe to the Crowing Crested Cobra of modern-day Africa – combines rooster with snake. Fear of gender transgression among chickens influenced the traditional Greek version of this monster deemed King of the Reptiles: it was believed the Basilisk materialized from the egg of a toad or snake that had been kept warm by a rooster (conversely, the Cockatrice, the medieval version of this abomination, emerged from an unusual egg laid by an old rooster and incubated for years by a toad or snake). Not just the venom but all aspects of the Basilisk – including his smell, bite and glance – were considered poisonous. For protection, travellers were advised to carry

The medieval Cockatrice emerged from an egg laid by a rooster and incubated by a toad or snake.

chickens with them because Basilisks could be killed if exposed to crowing cockerels – that is, when confronted by 'correct' chicken masculinity in the form of a crowing rather than laying rooster. Perhaps not surprisingly, reports of actual encounters with Basilisks are as scarce as hen's teeth; few have survived his stare to tell the tale. The last recorded sighting of a Basilisk occurred in 1587 in Warsaw when two little girls discovered one in the basement of a derelict house. Legend claims that the poisonous breath of this creature killed both children along with the nurse who came to find them, but the creature was finally dispatched by a condemned criminal sent into the house wearing a coat of mirrors.[50] Chaucer's *Canterbury Tales* mentions a 'Basilicok' and Leonardo da Vinci's *Bestiary* features a Cockatrice. The more recent coat of arms of Moscow incorporates a horseman spearing a Cockatrice.

In the region of southern Chile from which Araucana chickens hail, the Colo Colo hatches from an egg laid by an old rooster and is said to resemble a large mouse with the head of a cockerel. Feared because of his penchant for sucking saliva from sleeping humans, eventually causing their demise from fever, the Colo Colo hides in cracks and fissures in the house and can be heard to cry like a newborn baby.[51] One of the most popular Chilean football teams is named after this legendary creature.

The Korean equivalent of the Basilisk – the Kyer Yong (translated as chicken-dragon) – is sometimes depicted pulling chariots containing heroes or other legendary figures. Each year a ceremony is held in Ahmyong, South Korea, to celebrate Kyer Yong Mountain; folklore has it that a female Kyer Yong resides in a pool in this mountainous region and that women bathing in this pool obtain some of this creature's magical powers.[52]

Finally, across East and Central Africa, there have been numerous eyewitness accounts of the legendary Crowing Crested Cobra, said to resemble a snake with a bright red comb-like crest. Known by many names – *bubu*, *inkhomi* (the killer), *hongo*, *kovoko* – the Crowing Crested Cobra is around 20 feet (6 metres) long and lives almost exclusively on maggots obtained from decaying bodies. Both the male and the female sport the crest, but the male also possesses a pair of wattles. He has been heard to crow like a rooster when advancing on his mate, while the female cobra responds in turn by clucking seductively.[53]

4 Popular Chickens

Over the past century chickens have been represented in Western popular culture in several prominent ways reflecting the impact of shifting economic, social and political factors on human–chicken interactions. For example, portrayals of chickens in earlier twentieth-century fiction and film cast them as augurs of prosperity due to the wealth that new intensive farming practices promised to generate, while chicken characters post-1970s are more likely to be trivialized (the result of a broad devaluation of these birds following industrialization) or depicted as heroic 'underdogs' (helpful companions of troubled children or rebellious advocates of other fowl). In twenty-first-century media an image of vengeful chickenkind has also emerged, largely in response to environmental and animal welfare concerns regarding poultry farming.

Charlie Chaplin as the prophetic chicken in *Gold Rush* (1925). During the Cold War, Chaplin's left-leaning beliefs led to his exile in Switzerland where, coincidentally, he opened a chain of frozen-chicken restaurants.

PROPHETIC CHICKENS

Undoubtedly, the predominant part played by chickens in film and fiction has been that of food, a role that also situates the birds as important sources of income and wealth. In this capacity, chickens featured in American stories set between the two World Wars as the bearers of good fortune and better times ahead; their appearance predicted that the 1920s 'farm crisis' – which saw prices for agricultural crops drop dramatically – could be overcome by new intensive approaches to animal farming.[1] Charlie Chaplin's silent comedy *The Gold Rush* (1925), which he considered his masterpiece, stars the Little Tramp as a luckless prospector hunkered down for winter in the Yukon. In a hallucination brought on by hunger and cabin fever, Chaplin's companion Big Jim McKay (Mack Swain) sees the Little Tramp transform into a giant chicken – representing not just sustenance for them both, but also prosperity, for not long afterwards the friends discover gold and become rich.[2] After the Depression in America, 'better times' did indeed become associated with the rigorous industrialization of animals, especially of chickens for eggs and meat: battery farming was the new gold rush.

A similar account of the chicken as the bearer of prosperity occurs in one of Ray Bradbury's short stories set 'deep down in the empty soul of the depression in 1932'. A destitute family is travelling, like thousands of others in this era, west across America in search of employment. They come across The Inspired Chicken Motel (also the title of the story), which the father declares as 'a motel straight out of Revelations';[3] and, feeling some empathy for the large flock of chickens striving to survive in the hot and dusty Amarillo outback, they decide to stay the night. The landlady of this motel shares with them a special egg she keeps in a box:

> there were words written on this egg in white calcium outline, as if the nervous system of the chicken, moved by strange night talks that only it could hear, had lettered the shell in painful half-neat inscriptions. And the words we saw upon the egg were these: REST IN PEACE. PROSPERITY IS NEAR.[4]

This heralds the beginning of a chicken-inspired change of fortune for the family, which leaves the motel with a replenished sense of purpose and hope. Bradbury's story thus highlights the fantasy of the new industrial ideal that precipitated the move to intensive farming practices during the 1920s and '30s.

In the 1947 film *The Egg and I* (starring Claudette Colbert and Fred MacMurray, and based on the autobiographical novel by Betty McDonald), the success promised by the new mode of chicken farming is flavoured by hard work and sacrifice. Newlyweds Betty and Bob take on an abandoned chicken farm – Bob's dream – and have to adjust to rural life and responsibility for scores of hatched chicks. Portrayed in a light-hearted and optimistic manner, the 'bird-brain' antics of the chickens, combined with a neighbour's amorous advances towards Bob, help to popularize as well as sanitize the burgeoning business of commercial poultry farming.[5]

In sharp contrast to these depictions of chickens as bearers of wealth and good fortune, chickens also figured in early twentieth-century narratives as omens of troubled times ahead – prophets of dystopian futures. In these stories they are not personalized or anthropomorphized; instead, the existence of such uncanny or malevolent chickens warns of the negative consequences of human greed, as well as the problematic results of tampering with nature for scientific, technological and capitalist agendas. As early as 1904, H. G. Wells speculated about the risks of new technologies and the experimental manipulation of animals in *The Food of the Gods*.[6] Two 'gentlemen scientists', Bensington and Redwood, discover a food they name Herkleophorbia IV which induces rapid and substantial growth in organisms. Their test-subject chickens become gigantic predatory birds, escape, prey on children, and terrorize a village before being hunted down and killed.

Likewise, the Russian author Mikhail Bulgakov's science fiction novella *The Fatal Eggs* (1925) describes a scientific experiment that goes awry in Soviet Moscow.[7] A zoologist, Vladimir Persikov, discovers that amoebas left under a special ray of light show rapid binary fission. Persikov's discovery is appropriated by the authorities to revitalize Russia's defunct poultry business, which has been decimated by plague. Eggs are imported to be placed under the light-ray and replenish chicken numbers, but these are mixed up with a consignment of reptile eggs, and the result is an invasion of gigantic snakes and crocodiles.

THE TRIVIALIZED CHICKEN

The motif of 'the trivialized chicken' is very familiar to us today in the many jokes and idioms of abuse based on chickens (such as 'chicken-shit', 'dumb cluck' and 'choking the chicken'), and also in the term 'chicken' itself, now used as a synonym for 'coward'. It is also evident in cartoons and advertising where chickens tend to be anthropomorphized as dim-witted or silly. Such disparagement of chickens, however, is relatively new: as discussed in the previous chapter, across cultures and time roosters and hens have been admired for their vigilance, courage and loyalty to family or flock. The change in status of chickens is largely a result of twentieth-century industrialization and the consequent invisibility of actual chickens from the lives of modern-day consumers. As a result, we have come to rely on imagery and entertainment practices that all too often undervalue our historical relationships with chickens.

One example of a trivializing representation of chickens in media can be found in the successful on-line advertising manoeuvre launched in April 2004 for Burger King's new Tendercrisp™ chicken sandwich (the campaign's tagline: 'Get chicken just the way you like it'). Called Subservient Chicken, this publicity stunt involves an interactive website that allows users to command specific responses from a man dressed in a chicken costume. Subservient Chicken has been programmed to obey around 400 orders, including press-ups, dancing, spanking and watching television. If ordered to 'eat at McDonald's', he simulates being sick; if told to 'go vegetarian', he gives a thumbs-down.[8] Even though the figure of Subservient Chicken is clearly anthropomorphized, the banal acts he performs on command reinforce the idea of chickens as witless 'bird-brains'.

Another example occurs in a lucrative gimmick used in American casinos. Chicken Tic Tac Toe involves a live chicken in a specially designed cage, 'playing' against a human competitor. In fact, the chicken is responding to basic operant conditioning, having been trained to peck at certain cues to obtain food. The chickens win almost every game, although in 2004 Donald Trump reportedly trounced one of the chickens 'working' for his own casino in Palm Springs. A new Tic Tac Toe hen is traded for her predecessor every few months, and casino owners are advised on CasinoChicken.com 'after a set period of time (one to two months has worked often) [to] announce the retirement of the world-champion chicken (or fire it) and replace it with a group of a dozen "Apprentice Chickens"'. These novices are then advertised to draw in fascinated gamblers: 'Players will drag their friends into your casino just to put them through the fun humiliation of being "not as smart as a chicken".'[9]

The belittling of chickens occurs in other games in the United States. At Chicken Flying contests chickens are prodded with toilet plungers until they fall from heights of 12–22 feet (3.65–6.7 metres) and are forced to fly (an ability not well developed in chickens),[10] while to raise funds for 'Cowboys in Crisis' (injured rodeo riders) Chicken Roping contests are held for aspiring

young rodeo champs. At such events in states like Texas and
Colorado birds are subjected to lassoing by children, as well as
various forms of suspension and whipping by small ropes.[11]

Fruita's annual 'Run Like A Chicken With Your Head Cut
Off' race commemorates the tale of Miracle Mike the Headless
Chicken, another example of how the concept that birds' brains
are fundamentally lacking can be used to economic advantage.
In 1945 the anticipated routine slaughter for dinner of a five-
month-old Wyandotte rooster turned into an unexpected wind-
fall for Lloyd Olsen of Fruita, Colorado. Olsen's attempt to sever
the head of the chicken, later called Mike, resulted in part of
the bird's brain stem and left ear remaining intact. Thus Mike
lived on, blind and unable to eat, while Olsen fed him through
an eyedropper down his throat. The rooster survived for eighteen months as a sideshow chicken, appearing for audiences
from New York to San Diego for 25 cents per person. He choked

to death one night while on tour, when Lloyd misplaced the
eyedropper used to clear Mike's oesophagus.[12]

HEROES, REBELS AND UNDERDOGS

While the portrayal of chickens as frivolous and insignificant
creatures persists, this construction is increasingly challenged
in contemporary popular culture. For example, the 'heroic chick-
en' motif, which has emerged in the past couple of decades in
novels and on screen, depicts chickens as resourceful underdogs who take on the establishment, expose injustices and put
things right.

Lord's and Park's *Chicken Run* (2000), an animated production set in 1950s Yorkshire, is about a rebellion among chickens
destined for the farmer's new pie-making machine (the film's
tagline: 'This ain't no chick flick!').[13] The Tweedys' farm in *Chicken
Run* (a title that itself alludes to both captivity and escape) is

A scene from the
2000 Aardman
Animations film
Chicken Run.

like a POW camp, from which a clever hen, Ginger, frequently tries to flee. She is repeatedly caught, however, and often incarcerated in solitary confinement due to her unruly nature. When a rogue rooster called Rocket lands on the farm, he boasts that he can teach all the hens to fly the coop to a nearby chicken sanctuary, a feat eventually accomplished only with the aid of a pedal-powered flying machine (since chickens are poor flyers).

Disney's *Chicken Little* (2005) adapts the original fable of the same name (also known as *Chicken Licken* or *Henny Penny*), in which a hapless chick misinterprets a situation, believing catastrophe to be imminent and disturbing all those around him (a stereotype of the 'dumb chicken' or bird-brain). In the Disney version, Chicken Little really does end up saving the world: hit on the head by a curious object from above, he is at first derided for claiming that the sky is falling down – that is, until aliens arrive in the chick's home town, revealing that the device that fell from the sky belongs to them.[14]

The underdog status of chickens in contemporary Western culture means that they are often identified in film and fiction with marginalized human individuals or groups – usually with the aim of drawing attention to the plight of the latter. For example, Mark McNay's novel *Fresh* (2007) tells the story of working-class men employed at a Scottish chicken processing plant. These unskilled and impoverished men perform the kinds of unpleasant and unrewarding work that consumers refuse to think about. The chickens in McNay's novel are the lowest of the low, used literally as punching bags for those at the bottom of the human pile:

Chickens piled on his station like debts. He pulled them out with such force he could feel the hips dislocating. Come here ye bastards . . . [Sean] ducked and jabbed at the chickens as they passed him on the line. When a bird came he followed it down the line giving it left left-right. The bang on the fat breast had just enough give to make it feel like a human cheek. Sometimes the punch knocked a bit of fat out that looked like a tooth.[15]

Here, although the violence directed against the birds is graphic, it functions mainly to highlight Sean's own desperation.

Elsewhere, the association between chickens and marginalized humans may produce a sense of empathy for and between both parties. In Rob Levandoski's novel *Fresh Eggs* (2002), a compassionate relationship develops between battery hens and an eccentric child. Rhea Cassowary grows up on her father's million-bird battery farm, keeping her own free-ranging flock of Orpingtons with whom she shares a particularly special relationship. When she makes friends with the rebellious Miss Lucky Pants, a 'spent' battery hen who escapes the night-time culling of those in her shed, Rhea learns about the fates of the birds in her father's 'care'. The girl's embodied compassion for the incarcerated hens causes her to sprout feathers, first in easily hidden areas but then, when she becomes a teenager, all over her body. Freakish and isolated, Rhea takes solace in the company of her chickens, reading to them from the Persian classic *The Conference of the Birds*. Eventually, she escapes the factory farm, feigning her own death, and finds peace with a lover who adores her feathery uniqueness and who one day rescues – and reunites her with – the hens she left behind.[16]

This theme of the human oddball feeling at home with chickens occurs in other fiction for young adults. In *Chicken Boy* (2005) by Frances O'Roark Dowell, Tobin McCauley is a lonely boy, failing at school, his family torn apart by his mother's death. Then he meets Henry Otis and his brother Harrison, whose lives revolve around chickens. Believing that 'chickens are the center of the universe', Henry takes on a science project designed to discover the soul of the chicken. Henry explains to Tobin:

> 'Mr Peabody doesn't actually believe that chickens have inner lives. According to him, their brains are too small. But he hasn't spent enough time around real, live chickens. If he had, he'd know there's a lot going on inside those little heads of theirs . . . It was the chickens who make me realize life is right in front of us, but we ignore it . . . All my life I've been eating eggs, right? And eating stuff made with eggs? And, until recently, eating chicken. But until me and Harrison got our own chickens, I'd never seen one up close. I'd been reading about chickens in books since I was two. *Chicken Little*, *Henny Penny*, you name it. But I'd never experienced an actual chicken.'
>
> 'You weren't missing much' [Tobin tells Henry].
>
> Henry lean[s] in closer. 'Dude', he says, his voice almost a whisper. 'I was missing *everything*.'[17]

In joining the brothers' mission to discover the spirit of the chicken, Tobin resolves other interpersonal conflicts and deals with his grief over his mother's death. Like Tobin, chickens are under-appreciated: by learning to value their lives he gains new friends, both feathered and non-feathered, and realizes the importance of his own life.

Typically, in children's stories featuring animals, the child learns how to adapt to the adult world of responsibility and rationality. While this is accomplished with the help of animals, it tends to require the child's eventual detachment from a juvenile understanding of human–animal relations and the acquisition of a more 'realistic' one. Although novels involving relationships between children and chickens may follow this trajectory, they may also paint a more respectful picture of chickens as individual sentient beings: in *Chicken Boy*, through the boys' belief in chicken souls; and in *Fresh Eggs*, through the portrayal of Rhea as the central, most likeable and humane character in the story. Similarly, Clare Druce's *Minny's Dream* (2004) explores the friendship between a girl named Paula and Minny, a hen from a neighbour's battery farm. The dream Minny has is about her wild jungle ancestry and the natural lives of free chickens, while her nightmarish reality is that of a caged existence. In this case young readers are asked to consider the moral issues involved in rescuing or not rescuing this hen.[18]

THE AVENGING CHICKEN

While science-fiction novels such as Bulgakov's *The Fatal Eggs* and H. G. Wells's *Food of the Gods* warned readers a century ago of the dangers of interfering with nature and manipulating foodways to satisfy human gastronomic and financial greed, a similar admonition now comes in the form of 'the avenging chicken', an emergent motif in contemporary popular culture that draws on the idea of heroic chickens.

Known for its over-the-top sexploitation films, Troma Studios released in 2007 *Poultrygeist: Night of the Chicken Dead*, advertised as 'cinema's first chicken-zombie horror-comedy . . . with musical numbers!' When a fast-food chicken chain, called 'American Chicken Bunker', owned by Colonel Sanders's doppelgänger General Lee Roy, sets up a new shop on ancient Native American burial grounds, the spirits of chickens and people return from the dead to seek revenge. They kill those who work at the shop, and they either possess or inflict with deadly diarrhoea all those who consume the mass-produced chicken meat: 'this isn't just

General Lee Roy comes face to face with a mutant chicken in *Poultrygeist* (2007).

a mild case of salmonella . . . but something much more FOWL'.[19] Bad taste, 'superficially turgid and nihilistic' (in the words of one reviewer),[20] *Poultrygeist* does not advocate for chickens (or Native Americans, for that matter), nor is it serving the interests of intensive farming or corporate profit: another reviewer calls it 'a bold satirical comment on the chemical-industrial food complex'.[21] In fact, *Poultrygeist* exemplifies postmodern popular culture: borrowing from various genres and directly referencing themes from previous movies, it plays in a free-range fashion with controversial issues of mass meat production, fast food consumption, and the exploitation of humans and non-humans.

Chickens' revenge plots often cast Colonel Sanders look-alikes in the role of 'the bad guy'. Recording artist Moby accompanies *Disco Lies*, a song about betrayal, with a video in which a newly hatched chick witnesses other chickens being caged and slaughtered, and realizes that his kind has been deceived by people (and by one person in particular). Avoiding a similar fate, and growing up to become a human-sized and fashionable rooster, the chicken sets off to seek retribution. The traitor, a debauched and greedy man resembling the KFC Colonel, is pursued by the rooster until cornered at a chicken meat stall. There the villain is decapitated by the gallinaceous hero, who in the final scene enjoys a victory platter of human legs drizzled with gravy and set upon a fine looking salad.[22]

'CHICKS' AND 'CHOOKS'

> She's no spring chicken: she's on the wrong side of thirty, if she be a day.
> Jonathan Swift (1667–1745)[23]

Like children, women have enjoyed close associations with chickens – and not just in literature and film, but also in particular cultural contexts. This may be partly due to the common marginalization of women and chickens in many societies – noted, for example, in slang such as 'chicken dinner' (an attractive woman), 'chicken ranch' (a brothel) and 'chicken hawk' (an older man pursing younger women for sex); and also because chicken meat itself has tended to be feminized (like other white meats) – viewed as weaker or softer than masculinized forms of red meat such as beef.[24] Across cultures women have also often been closest to chickens, raising and caring for them, selling, killing, cooking and consuming them;[25] today in countries

such as Uganda, Afghanistan and Pakistan women remain at the forefront of entrepreneurial chicken farming, setting up small egg-laying flocks in urban environments to supplement household incomes.[26]

African American women have a particularly significant historical relationship to both living chickens and to chicken meat. Since the era of slavery in the United States, chickens have provided sustenance and income for African American families, influenced their culture, and enabled black women to define and empower themselves against racist stereotypes. It is also the case, however, that racist propaganda in the US has drawn on African Americans' traditional proximity to chickens to portray them in compromising situations, for example, as chicken thieves or victims of chicken attacks. Deprecating and 'mammified' images of black women cooking chickens, in which the women are either desexualized or eroticized, occur in older cartoons, posters and literature, as well as contemporary stand-up comedy, film and television. Alongside such demeaning imagery, an alternative strand of representation is present in black women's own accounts of their culinary and trade successes involving chickens. In Gordonsville, Virginia, for instance, the local entrepreneurial black women known as 'waiter carriers' (or chicken vendors), who sold their food at train stations a century ago, established both status and income through their business.[27]

Chickens also have particular gender, age and class associations in Australian and New Zealand settler cultures. Judith Brett contends that 'chooks' (pronounced like 'books'), as these birds are referred to 'down under', represent the fortitude of the pioneering spirit. They are inventive and determined, often portrayed as misfits or rebels and viewed by Australasians as tough little heroes: 'Scratching out an existence from unyielding ground, collapsing into a flap when danger threatens, the

Waiter carriers on Gordonsville railway station platform selling fried chicken in the late 1800s.

chook not the desert haunts our dreams.'[28] The very word 'chook' embodies the cultural uniqueness of Australia and New Zealand, and is proudly employed as a conscious marker of working-class identity and (supposed) indifference to class.

Older women in both countries may also be referred to as chooks. In this case the word has negative connotations, as a mild form of insult that expresses the widespread disparagement of older age in women. Yet some recent writers have taken up the specifically Antipodean connection between women and chooks in a more positive way, employing the motif of the underdog (or 'underchook') in narratives that interweave the lives of women and hens. In *Book Book* (2003) the New Zealand author Fiona Farrell's protagonist is captivated, as a child, by stories about chickens. She carries *The Little Red Hen* wherever she goes, entranced by the industrious chicken who made bread by herself while her lazier comrades refused to help. The little girl also learns from the mistakes of *Chicken Licken*: 'You must not panic, nor imagine that things are worse than they are. You must be sensible, you must be resourceful. You must manage on your own and make your own bread. Those were the lessons taught by hens.'[29]

5 *Gallus graphicus*

Art involving chickens has taken many forms since the oldest documented figurines of a cock and hen were found in the Indus River valley (today's Pakistan), which date to around 3,000 years ago. As noted by one historian of chicken art, 'There aren't too many artistic "-isms" in which the chicken isn't'. From realism to naturalism, impressionism to cubism, Dadaism to surrealism, the chicken has inspired the masters, and a surprising number of great artists have instances of chicken art in their body of work.'[1]

FAMILIAR CHICKENS

Images of roosters, symbolic of masculinity and virility, are common on ancient Greek artefacts, with cockfighting scenes prominent on vases and bowls. The love affair between Zeus (god of the sky and thunder in Greek mythology) and Ganymede (a Trojan prince) also appears on such vessels, often showing the young man carrying a cockerel given to him as a courtship gift by his divine admirer. In ancient Rome chicken motifs tended to appear on more practical objects, such as coins and lamps.

Following the fall of Rome, chickens receded from European art (with the exception of bestiaries) for many centuries, until

Greek vase showing Ganymede with rooster, *c.* 485 BC.

re-emerging at the end of the Middle Ages. From the fifteenth century to the seventeenth they commonly appeared in the works of Dutch painters such as Rembrandt, van Horst and Melchior d'Hondecoeter (1636–1695), who was renowned for the liveliness of the birds he painted.[2]

The rooster has, of course, held a special place in French national culture, in part because of the similarity between the Latin words for cock (*Gallus*) and for the inhabitants of Gaul (*gallicus*). As Christian symbols of vigilance, rooster motifs appeared in French churches during the medieval period, and from the sixteenth century coins sometimes showed cockerels alongside the king of France. The French Revolution saw the proliferation of rooster imagery on official emblems, but Napoleon reputedly regarded chickens as weak, preferring the eagle as a representation of France. The rooster regained status in France in the nineteenth century, however, figuring on the uniforms of the National Guard, replacing the fleur-de-lis as the national emblem, and gracing the Great Seal of France.[3]

Roosters have also featured in French high and popular art. In the summer of 1874 'a momentous occasion happened in the history of chicken art', when Auguste Renoir and Eduoard Manet painted Claude Monet's family in their garden at Argenteuil; in both paintings Madame Monet is shown sitting with her son on the lawn with chickens nearby, acting as 'regal bystanders and observers of rural life'.[4] Meanwhile, the French cartoonist known as J. J. Grandville, famous for illustrating such works as *Gulliver's Travels* and *The Adventures of Robinson Crusoe*, was portraying human bodies with the heads of hens and roosters in his satirical illustrations.[5]

Chickens were prominent in various styles of art throughout Europe. The Spanish artist Francisco de Goya (1746–1828) featured a dead cockerel in his series of still-life paintings, many

In this stylized image from a First World War poster, the chicken is depicted as a soldier who aids the war effort.

of which show slaughtered animals awaiting dismemberment and consumption, while the German Expressionist Franz Marc (1880–1916) produced an early drawing of a very life-like hen peering downwards, writing in 1908 that his purpose was to 'animalize art', to challenge the convention of portraying animals as objects in landscapes by painting with a view to 'seeing *with* the animal'.[6]

In the mid-1950s René Magritte (1898–1967) painted *Variation of Sadness* (*Variante de la tristesse*), which shows a chicken looking at an egg in an eggcup, as if perplexed by the fate of her chick,

René Magritte,
*Variante de la
Tristesse* (Variation
of Sadness),
1955/7, oil
on canvas.

while his fellow surrealist Paul Klee (1879–1940) created a similar but darker image of a hen peering at an opened egg exposing yolk and white, a giant red exclamation mark above the eggcup.

Pablo Picasso (1881–1973) also produced a number of paintings in which roosters, and occasionally hens, were the focus. He sketched chickens in charcoal and chalk from 1896 until 1921, when he created his first major oil painting of a rooster, *Dog and Cock*, as part of his synthetic Cubist period. One of Picasso's most important works is the *Cock of the Liberation* (1944), completed while the artist was living in Nazi-occupied France, and thought to represent the fighting spirit of the French.[7]

Marc Chagall (1887–1985), of Russian and Jewish descent, demonstrated an admiration for roosters throughout his artistic career. These birds frequently appear in his works as vibrant central motifs or accompanying prominent human characters. Chagall portrays the rooster as a guardian and friend, protecting lovers under his wing, pulling a child through the air on a sled and leading a bride and groom from the Eiffel Tower to their wedding. Chagall's love of chickens was perhaps a legacy of his grandfather's occupation as a butcher, for in his autobiography the artist admits that witnessing animal slaughter as a child left a deep imprint.[8]

Chagall's roosters suggest that the place of chickens in art cannot be understood in purely aesthetic or symbolic terms, divorced from their place in wider society. Indeed, animals have always been crucial to the ways in which we come to understand and form images of ourselves as humans.[9] Art also shows, and reflects on, the practices that have historically dominated human–animal relations, such as agriculture and hunting.[10] The changing roles of chickens in art tell us about changes in human life, and changes in the human treatment of animals.

In Victorian Britain, for example, the so-called hen craze led to an explosion of chicken breeds, fanciers' clubs and formal competitions. Chicken paintings of the Victorian era therefore aimed for life-like accuracy in depictions of the birds, concentrating on portrayals of hens or roosters in romantic pastoral settings such as farmyards and meadows.[11] During this period, illustrators of chickens were also in demand for the official documents and reference books of the poultry fanciers' clubs. J. W. Ludlow (1840–1916) was responsible for the pictures in several such texts including Lewis Wright's *The Illustrated Book of Poultry* (1873) and *Cassell's Poultry Book* (1912), while Harrison William Weir (1824–1906) drew for poultry magazines and designed his

overleaf:
Marc Chagall,
The Cock, 1950,
lithograph.

Gustav Klimt's
*Garden Path with
Chickens*, 1916,
oil on canvas

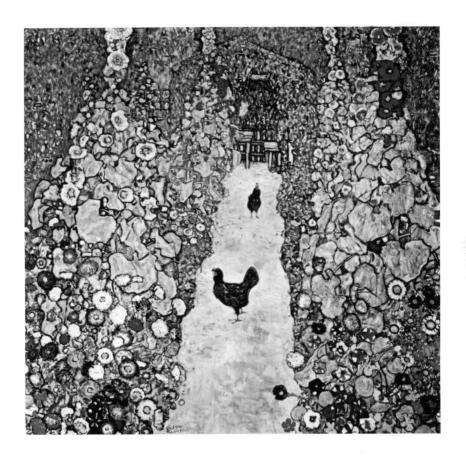

own heavily illustrated tome called *Our Poultry* (1873). Around this time across the Atlantic A. O. Schilling (1882–1958), described as 'one of America's foremost chicken artists', provided the pictures for the *American Standard of Perfection*.[12]

The Kansas-born John Steuart Curry (1897–1946) and Grant Wood (1891–1942) from Iowa are most closely associated with the art movement known as Regionalism, which favoured depictions of American Midwest rural themes in a deliberate move against European art traditions. Chickens feature strongly in their works, promoting the stoicism and self-reliance of country

Grant Wood, *Appraisal*, 1931, oil on composition board.

life. Wood's *Appraisal* (1931) depicts two women interacting, one a well-groomed fur-clad 'lady' from the city, the other a 'self-reliant farm woman' cradling a Plymouth Rock rooster and 'standing politely firm against intrusion from the city'. This painting has been read as a critique of the technologization and materialism associated with urban lifestyles compared with the practical and more down-to-earth existence of those in the declining agrarian sector between the World Wars.[13]

Curry's *Hen and Hawk* (1934) shows a mother hen fighting off a hawk who has attacked her chicks, while his later mural, *The Social Benefits of Biochemical Research: Farm Stock, 1941–43*, which he created for the biochemistry building at the University of Wisconsin Madison,[14] shows various farmed animals frolicking happily in front of a huge barn-like structure, while hens (both free-ranging and caged) are guarded by a crowing rooster – this latter work implying that both rural society and farmed animals will be advantaged by new agricultural technologies.

In North America chickens have also been popular subjects of 'folk art', those modes of creative and often practical works generated outside formal institutions and not adhering to historical assumptions or ideals about 'art'. The self-taught Pennsylvanian artist Ben Austrian (1870–1921), nicknamed 'The Chicken Painter of Berks County' and celebrated in particular for his paintings of mother hens and chicks, informed a newspaper reporter in 1900: 'I paint chickens because I love them.'[15] One of Austrian's images of a newborn chick (*Hasn't Scratched Yet*) became the trademark of Bon Ami Cleanser more than a century ago.

The African Americans Bill Traylor (1854–1947) and Jimmy Lee Sudduth (1910–1997) are known for their 'outsider' folk art, so called because such works stand apart from the mainstream and outside dominant aesthetic traditions. Born into slavery in Alabama, Traylor did not start drawing until he was in his

Jimmy Lee Sudduth, *One Big Hen*, 1985, earth pigments and house paint on plywood board.

eighties, producing more than 800 pictures – mostly of animals and people from his childhood – until his death at the age of 95. His paintings of chickens employ vivid colours such as yellow and blue to portray the birds' exuberant passion for life. A fellow Alabama resident, Jimmy Lee Sudduth, became something of a legend for his 'mud paintings', comprising pigments obtained from dirt, leaves, berries and other natural elements. His work includes numerous bold and earthy paintings of hens and roosters.[16]

Chicken folk art continues to be popular in the USA. In the Californian town of Fair Oaks, inhabited by feral chickens for

30 years, Carol Rhodes-Wittich has established the Museum of the Chicken, 'which honors all things chicken'. The town also hosts an Annual Chicken Art Show and Decorate a Chicken Contest, the purpose of which is to dress up a plain wooden chicken as outrageously as possible, and an annual Chicken Festival, attended by some 8,000 people, which allows the public to select winners from the decorated chickens.

Roller Rooster – contemporary chicken folk art by Cappi Phillips.

An early 20th-
century print of a
rooster and a hen
by Ohara Koson.

Asian art is also replete with images of chickens, especially cocks. In China, the rooster, one of the twelve animals of the zodiac, represents courage, strength, honesty, pride and vigilance. The rooster heralds good luck and prosperity and is believed to be able to exorcize negative forces and warn of dangers. In Chinese folk art, two popular ancient motifs depict a child clasping a rooster, and a cockerel with a fish in its beak. Various kinds of fortune and power are signified by the different number of roosters shown, and by the stance or activity undertaken by the birds. For instance, five roosters drawn together represents the importance of fives (such as the 'five martial arts' and the 'five concepts'), while an image of a rooster crowing indicates 'scholarly honor or official rank'.[17]

Wild roosters are similarly respected birds in Japan, free to roam throughout Shinto temples. A Kyoto native, Ito Jakuchu (1716–1800), is perhaps the most celebrated of Japanese chicken artists. He kept fowl at his home in order to observe them more closely for his art.[18] Most of Jakuchu's works appear on hanging scrolls, and reflect both Eastern and Western artistic influences, as well as Zen Buddhism.

In Korea there is a private museum that pays homage to wild and domesticated fowl. The Seoul Museum of Chicken Art, which opened in 2006, houses an array of artefacts and crafts focusing on chickens, from antiques to contemporary sculptures and paintings. Visitors have the opportunity to learn about the history of chicken–human interactions, as well as various forms of chicken art from both Eastern and Western traditions and contexts; and exhibitions are changed each season in keeping with the impact on chickens of spring, summer, autumn and winter.[19]

The diverse artworks surveyed so far can be characterized as *familiar* renditions of the chicken, that is, representations of commonly shared ideas about 'chicken-ness' and what chickens symbolize within their particular cultural contexts. In other artworks, however, the image or idea of the chicken challenges the viewer by evoking a sense of the *unfamiliar*. Such works may disrupt conventions by associating chicken motifs with other, often less sanguine, aspects of chickens' lives or appearances, or by using chickens as metaphors for human concerns. The San Francisco-based artist Doug Argue, in an untitled work from 1994, depicts scores of layer hens in battery cages, evoking the sense of defamiliarization connected with contemporary post-1960s art. The sheer size of this painting (12 by 18 ft, or 3.65 by 5.48 metres) makes it a challenge to the senses, while its subject-matter, a huge battery farm in which the details of individual

Doug Argue,
Untitled, 1994,
oil on canvas.

chickens are apparent, is liable to make the viewer uncomfortable. Arguc's mural, inspired by a Kafka short story in which a dog considers his life in relation to how he gets his food, represents for the artist the changes that have occurred in housing for humans and for 'food animals' in modernity. More recently, the artist has completed a series of paintings of individual chicken subjects, each giant chicken representing a bird from the first mural projected to monumental scale. This process reverses the reduction of individual hens to disposable components in a vast battery of food production.[20]

Sometimes chickens are used to draw attention to political or social issues affecting humans. Ezrom Legae (1938–1999) was a South African resistance artist whose political sculptures and pictures protested against the injustices of apartheid. In his famous *Chicken Series* (1978), created after the death of his friend Bantu Steve Biko, Legae used charcoal and pencil to draw intricate and disturbing images of chickens who were in pain and

Inside a modern battery farm.

torment, but nevertheless able to transcend their suffering and metamorphose into beings more powerful. It is suggested that Legae drew on 'the common ordinariness of the chicken to represent the impossibility of destroying the spirit of popular protest'.[21] Legae himself stated that he 'used the chicken as a symbol of the black people of this country, because the chicken is the domestic bird. Now, one can maim a chicken by pulling out his feathers; one can crucify him and even kill him; but beware – there will always be another egg and always another chicken.'[22]

In some contemporary works a sense of transgression or estrangement is generated through the exploitation of actual birds, which may involve the use of chicken skin, feathers, claws, meat or whole carcasses. In industrialized societies we tend to be distanced from the creatures we consume; slaughter is hidden from the average urban omnivore (although this is changing somewhat with the emergence of popular New Carnivore shows such as *Jamie's Fowl Dinners*, one episode of which focused on Jamie Oliver's graphic on-stage slaughter of hens for his next dish).[23] Chickens and other farmed animals can be used in performative or experimental art to render more visible the increasing power of science, technology, agribusiness and consumerism. Koen Vanmechelen's *Cosmopolitan Chicken Project*, for example, comprises a ten-year 'attempt to create and manipulate scores of chicken breeds from all over the world into a new species, a universal chicken or Superbastard'. The Belgian artist contends that 'the chicken and the egg are a metaphor for the human race and art'.[24] Works of this nature may aim to disturb complacency about the routine exploitation of chickens, but at the same time chickens are used as 'tools' by the artists themselves. A more ruthless example is provided by the Chinese performance artist Sheng Qi, who in his 1990s show *Universal Happy Brand Chicken* began by

Pinar Yolacan,
Perishable Art series,
2004.

caressing and kissing several live chickens before injecting them with chemicals, stabbing and dismembering the birds and finally urinating on their carcasses. This work was supposedly designed to draw attention to cultural differences in the treatment of animals,[25] but it may also be read as a lesson in the ubiquity of animal abuse across cultures.

The move to employ chickens and other animals – dead or alive – in contemporary art is obviously not always motivated by pro-animal sentiment or animal advocacy. The bodies of dead chickens are also fundamental to the work of Turkish-born New York-based Pinar Yolacan. Her *Perishable Art* series (2004) involved the creation of Victorian-style fashion garments, embellished with ruffles, frills and fancy collars made from the parts of recently killed chickens and modelled by women in their seventies. In this case eviscerated chickens were used by the artist in her disruption of the traditions of Western portraiture and in her subversion of ideas about ageing, death, temporality and impermanence.[26]

Art using dead chickens or parts of chickens to provide metaphors for human death, loss or misery is ultimately anthropocentric – it centralizes human concerns and interests. The bodies of dismembered chickens, however, may also influence art created in the name of animal activism. In this context, rather than serving the expression of human experiences and perspectives, the transgressive practice draws attention to the fate of the birds themselves.

New Zealand-based Angela Singer works with discarded carcasses to disrupt the taken-for-granted ways in which animals are raised and slaughtered for food. For one exhibition in Auckland, Singer created a piece emphasizing the 'throw-away' parts of chickens. *Chicken Kitchen Curtain* involved salvaging chicken feet cast away by a local butchery, attaching these to a sheet of latex and powder, and painting the 'curtain' the orange colour of the fake chicken seasoning sprinkled on French fries. Singer explains:

> The chicken kitchen curtain hung in a small space, slightly out from the wall – the chicken claws stuck out and caught on people's clothing when they walked too close. At the opening of the show I saw people react quite angrily, and some were revolted when they realized what the curtain was made of. There was another show opening in the gallery the same night. The finger food included chicken meat so people were coming out of that show, and walking into my show eating chicken and getting upset about chicken claws![27]

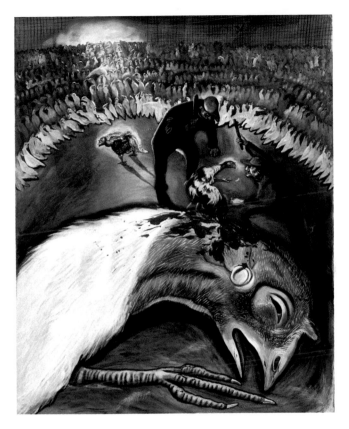

Sue Coe, *No Humane Death for Poultry*, 2000.

The critique of capitalism and animal exploitation is also central to the art of the New York-based, British-born animal activist and self-described 'visual journalist' Sue Coe. Her distinctive and provocative works are meant to shock: they portray explicit cruelty to animals occurring in laboratories, slaughterhouses, hatcheries, farms and the entertainment industry. Coe does not shy away from graphic scenes of chicken processing plants, or intensive farming practices such as de-beaking and the mass killing of male chicks; in fact, her art is directly influenced by witnessing first hand such routine procedures in factory farming situations. Such a focus on pain and death is common in animal art that aims to show how the distinction between humans and other animals is symbolically constructed and practically enacted.

Mary Britton Clouse of Minneapolis is another artist dedicated to undermining negative assumptions about domestic fowl. Her recent works include photographic portraits of real birds that she has personally rescued and rehabilitated. Blurring the line between portrait and self-portrait, Clouse's art is fundamentally practical in its formation: the images are obtained from fidgety birds being snapped for the adoption pages of her chicken sanctuary's website (in fact, Clouse holds the birds with one hand

Mary Britton
Clouse, *Nemo:
Portrait/Self-portrait*
(2005).

132

while taking pictures with the other). The arresting image of *Nemo* was the product of just such a process, whereby the faces of human and chicken were momentarily, without posing and entirely accidentally, captured by the camera, in perfect unison. Such images demonstrate the power of some animal art to challenge taken-for-granted notions about coherent species identity, whether of humans or of other animals.

A fellow American artist, Nicolas Lampert, uses transgressive political art and street activism to critique unrestrained industrialization and consumerism. He highlights the complicity of technology and agribusiness in hierarchical systems, including the dominance by humans of animals. In his *Meatscapes* series, beginning in 2000, Lampert inserts images of chicken meat sourced from old cookbooks – raw or cooked, cuts or whole carcasses – into romantic settings, rural scenes and well-known tourist destinations from travel catalogues. These collages subvert traditional landscape painting and are flavoured by kitsch humour:

The *Meatscapes* are meant to be absurd, but there is a serious side to them as well. The collages, which act more as staged photographs, are meant to ask questions about the larger implications of being disconnected from our food source. The people in the images are purposely nonchalant about the massive piles of meat, which echo the general lack of understanding of the environmental impact of large-scale industrial production of meat.[28]

In July 2007 Lampert and Micaela O'Herlihy placed a gigantic polystyrene foam broiler chick carcass in the parking space of the Big Lots supermarket in Milwaukee. This unavoidable eyesore, named *Attention Chicken!*, was intended to confront consumers as they entered the shop, perhaps on their way to purchase chicken

meat itself. *Attention Chicken!* has since mysteriously appeared in various locations across America – on a beach, in the woods, and on a corner in Pittsburgh prior to the arrival of a festival parade.

Although the realities it deals with are grim, activist art can be humorous, as Lampert's work shows. Tasmania-based Yvette Watt strategically employs humour and anthropomorphism to promote an empathetic association between the viewer and the animal depicted. Influenced by the physiognomic studies of human and animal faces of the seventeenth-century French painter Charles Le Brun, who believed that facial congruities between human and non-human animals were indicative of certain emotional or behavioural types, Watt has created paintings and photographs in which her own face appears, with her eyes replaced by those of a farmed animal such as a chicken. The resulting hybrid images, for example those in Watt's series *Second Sight*, invite the viewer to see the world from the bird's perspective.

Watt has also painted a series of paintings called *Offerings*, based on real-life farm animals. *Sally*, one portrait in this series, is based on a rescued battery hen now living on a farm sanctuary. Her image is painted onto a white linen tea towel in Watt's own blood. Because this medium quickly changes to a sepia colour as it dries, the artist is able to ensure that the viewer's initial engagement is with the image rather than the more sensational connotations of its medium. In Watt's words:

> The intention is that, on discovering the nature of the painting medium used, the viewer will be caused to consider the matter of these animals as flesh and blood – and hence as meat. As such it was essential that the blood used was *my* blood, as I see these works as gestures of solidarity with those animals that are killed in their billions for meat; as a kind of offering, a symbolic giving up of my blood, a

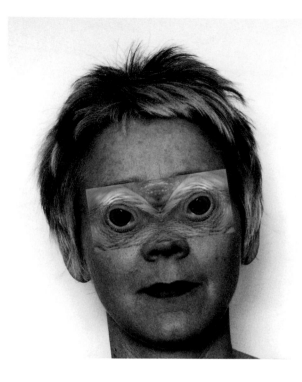

Yvette Watt,
*Second Sight
(Chicken)*, 2007/8,
giclée print on
photo rag paper.

recognition of the spilling of the blood of these animals for meat production and of the fact that their blood stains the kitchens of most homes.[29]

The previous chapters, in surveying the place of chickens in folklore, popular culture and art, have brought us all the way from the cosmological mythologies of ancient cultures to the mundane realities of the contemporary kitchen. In the process they have also brought us inexorably – as our species has brought *Gallus gallus* – to the domain of the modern meat industry, which is the subject of the next chapter.

opposite:
Yvette Watt, *Sally*
(from *Offerings*
series), 2007,
artist's blood on
linen tea towel.

136

THE POULTRY YARD.

6 Meat Chicks and Egg Machines

Worldwide, more than 50 billion chickens are killed for meat each year. In the time required to read this page 16,000 will be killed in the USA alone. America kills around 23 million broiler (or meat) chickens per day, 8–10 billion per year.[1] In the UK over 860 million broiler chickens and 30 million 'end-of-lay' hens are killed annually. Australia kills 500 million broiler chickens each year for meat, having raised 96 per cent of them in intensive systems, while 11 million battery hens produce 93 per cent of the nation's eggs.[2] Australians refer to the modern poultry industry as 'technology's child'; this chapter shows why.[3]

In just over a century chickens have been transformed from birds revered for their bravery, fortitude and devotion to parenthood, to the least respected and most manipulated beings on the planet. The term 'chicken' now symbolizes cowardice, and the hen, whose love for her chicks was once so admired, has become a dispensable egg-making machine. Instead of a natural lifespan of up to twelve years, the typical farmed chick today will live for about six weeks, never having experienced sunshine, rain or grass. In modern societies these birds have become de-natured, de-personalized and even de-animalized.

The modern enslavement of chickens is a very Western and capitalist tale. Its origins are in America, specifically the East

The Poultry Yard, c. 1869, lithographic print.

Coast region of Delmarva, where the broiler industry arose, and Petaluma in California, where battery farming for eggs began. Breakthroughs in animal husbandry converged with technological and commercial developments around the beginning of the twentieth century radically to alter the role and status of the chicken. At each stage of research and development the chicken's natural proclivities were subordinated or exploited to produce the industrial 'utility bird' of today.

Young poultry vendors in Cincinnati, Ohio, in 1908.

Before the First World War, chickens were familiar residents of small farms and suburban backyards. They were mainly cared for by women and children, their eggs sold to supplement household income. Meat came from young cockerels ('spring chickens', the forerunners of today's 'broilers', 'roasters' and 'fryers') and older 'end-of-lay' hens. Under natural conditions (and depending on breed) a hen will lay as few as 30 eggs a year, mostly in the spring. Eggs were therefore in demand over off-season periods, while prices fell when they were more available. These fluctuations rendered eggs and chicken meat delicacies.

Chickens at Rousham House, Oxfordshire. Many of today's amateur chicken-keepers prefer to go back to pre-factory techniques and let their chickens free-range.

The development of new farming technologies and practices removed the seasonal limitations on the poultry business. The first major inventions were the incubator, designed by a resident

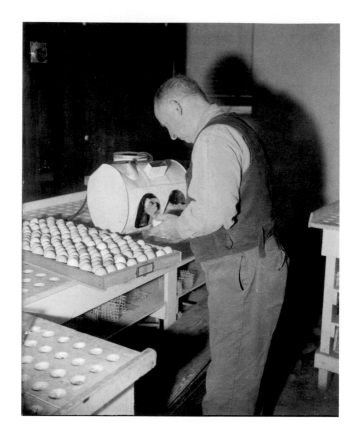

of Petaluma, California, Lyman Byce, in 1879, and the colony
brooder. The former offered automatic ventilation, controlled
temperature and mechanized turning of eggs, permitting the
artificial incubation of vast numbers of eggs at once, while the
latter allowed 300–1,000 hatched chicks to be raised together
under stoves heated by kerosene or coal until their feathers
grew.[4] Both apparatuses separated chicks from their mother
and the natural environment.

Meanwhile, scientists were examining all aspects of chicken histology, physiology, anatomy, genetics, nutrition and pathology. It is no coincidence that the first creature to have a full genome map was the chicken: *Gallus* is the most studied species in the world.[5] Every biological feature was experimented on with the aim of rendering chickens more serviceable to humans. From the early twentieth century the chick's ability to obtain sustenance from the yolk sac immediately after hatching was exploited in the transportation of thousands of newly hatched chicks across large distances – a development augmented by the invention of the automobile and refrigeration. The cessation of laying by hens when daylight hours shortened – a behaviour considered most

Demonstration of internal workings of the hen given by H. L. Shrader of the US Department of Agriculture. This model with mechanical organs was shipped to the World's Poultry Congress in London, 1939.

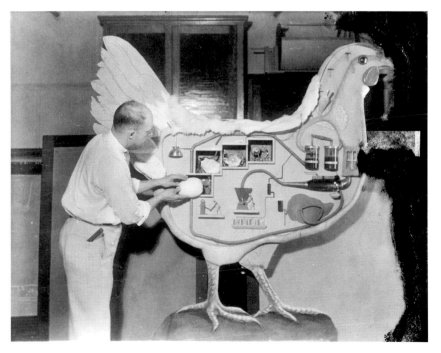

Poultry farm,
Manzanar
Relocation Center,
California, 1943.

inconvenient and uneconomical by farmers – was foiled through
the introduction of artificial lighting in henhouses.

Specialized breeding programmes also accelerated in the early
twentieth century. The 'artificial evolution' of chickens through
selective breeding had already begun during the late nineteenth-
century 'hen craze', when chickens of different varieties were all

the rage. The careful monitoring by poultry associations of the criteria for fancy breeds produced a more portentous outcome because it demonstrated that certain chickens were superior layers, while others were better to eat. This resulted in the separation of egg farming from the farming of chickens for meat.

The most profound change, however, was the confinement of thousands of chickens indoors for the purposes of management and control. As farmers began to keep more and more birds, monitoring them became difficult and time-consuming, so they were moved into barns or cages for easier handling. This crucial step in the process of industrialization was only partially successful in the early twentieth century, since chickens developed leg weakness away from sunlight. Once researchers in the 1920s solved this problem by the addition of Vitamin D to the food of confined chickens, the lucrative business of specialized poultry feed production began.[6]

Prior to battery farming white hens had not been favoured as backyard birds because they attracted predators. Customers preferred their white eggshells, however, and once hens were contained indoors the colour of their plumage no longer mattered. The White Leghorn, a naturally generous egg-layer, became the focus of commercial breeding programmes, and the Petaluma district in central California, where the incubator was invented, began farming White Leghorns en masse in the 1920s, quickly earning the label of the 'Nation's Egg Basket'.[7]

One of the earliest endorsements of battery farming appeared in 1927 in the *National Geographic* magazine. In 'America's Debt to the Hen', Harry Lewis foretold the shift in status of the egg-laying hen, from a 'universal favorite [who had] been a companion of man in the onward march of civilization' to a vital component in the faceless machinery of the new 'factory farm'.[8] A few years later G. W. Wrentmore, the British author of *The Battery System of*

Poultry Keeping (1931), presented this new form of managing layer hens as a patriotic necessity:

> [Americans] soon realized that being able to keep the birds in confinement was the poultry farmer's dream come true ... Why should not the whole of England follow on the Petaluma lines ... Let us produce ... BE BRITISH. GET DOWN TO IT. DO YOUR BIT.[9]

Wrentmore went on to claim that the new farming houses, the 'arrangement [of which] suggested a sort of battery construction', were advantageous to the birds themselves: while they might hamper the birds' natural activities, their safety from 'children throwing stones' was assured.[10]

These technological 'advances', however, did not go unopposed: objections from the Society for the Prevention of Cruelty to Animals were accompanied by those of some chicken fanciers who felt that the nature of birds, and the 'art' of chicken breeding, were being violated. Others lamented the ways in which the inherent behaviours of chickens were being denied by industrial farming.

A hundred years ago, when chickens were kept predominantly for eggs, their free-ranging resulted in strong muscles and tougher meat so they were seldom raised to be eaten. Those birds destined for the dinner table were often 'conditioned' (force-fed) a few days prior to slaughter, and the form of carcass preparation known as New York Dressing was common.[11] This involved plucking chickens and draining their blood, but selling them with heads, feet and internal organs intact. Getting a New York Dressed bird ready for dinner was time-consuming because it required decapitation and evisceration, tasks usually undertaken by the homemaker. An increasing demand for chicken meat after the First World War, fuelled by a populace who enjoyed eating outside the home more often and were keen to try different foods, motivated the poultry industry to investigate more convenient and lucrative alternatives.

In the 1920s American chicken meat production still remained largely the domain of farmers' wives in the Midwest and South. Raising flocks was a feminized enterprise before the mid-century; for men, it was even considered a 'disgrace to be caught in the chicken house'.[12] Credit for continuous year-round production of broiler chickens goes largely to Mrs Wilmer Steele of Delaware, who in the spring of 1923 was mistakenly given 500 chicks by a local hatchery instead of the 50 she had ordered to replenish a culled-down flock of layer hens. Rather than returning the extra chicks, Mrs Steele opted to build a bigger chicken house and raise them all for meat. Five months later she sold the surviving 387 chickens, weighing around 2 lb each, to a middle man who transported them to the surrounding cities. By 1926 Mrs Steele was 'growing' 10,000 birds at a time and others were noticing her success. So it was that on the Delmarva (Delaware-Maryland-Virginia) Peninsula, close to major markets such as Baltimore, Philadelphia and New York, commercial broiler farming first flourished. Within a few years 7 million broiler chickens were being raised annually in this region,[13] and the Republican Herbert Hoover felt confident enough during his presidential campaign of 1928 to promise Americans 'a chicken in every pot'.[14]

The Delmarva Peninsula dominated the American broiler industry until the Second World War, when supply to the general population from this area was frozen to meet the army's needs, and the wider US market was lost to regions further south.[15] Owing to the masses of broiler chickens now being farmed, it was considered (as with layer hens) no longer efficient to have flocks roaming freely at risk from predators, weather and disease. The

Judy Chicago,
A Chicken in Every Pot, 1998, painting, needlepoint, appliqué and embroidery.

A Chicken In Every Pot

modern broiler house was designed to provide indoor captivity for thousands of birds, and its use increased noticeably from the late 1940s (meanwhile, to deal with the risks of keeping so many birds in such close proximity, scientists worked on disease prevention, vaccination and antibiotic administration). Profits were dependent on efficiency, equated with growing the heaviest

148

chickens within the shortest time. Combining appetite stimulants in high-energy foods with restrictions on the birds' mobility ensured the rapid development of monstrous baby chickens, who reached their death-weight well before they were sexually mature – thereby removing the need to separate female and male chicks, although some farms raise sexed chicks separately to ensure uniform weights in flocks.

Feeding programmes were also a high priority for the broiler industry. Genetic improvements in progeny testing, along with experiments involving breed and strain crosses, helped produce birds who grew larger and more rapidly. A national contest named Chicken of Tomorrow, held in 1948 at the University of Delaware and supported by poultry companies, university researchers and retailers, encouraged the public's interest in 'better broiler' breeds and increased the popularity of chicken flesh as an 'everyday meat' in America.[16] The contest's objective was to find the ideal chicken, imagined to be like a broad-breasted turkey with plentiful flesh. Each contestant was required to send two cases of fertilized eggs to a hatchery in Maryland; these were hatched and reared in the university's experimental unit in 40 separate pens. After twelve weeks the chicks were weighed and placed in feeding batteries for three and a half days, after which they were killed and their carcasses judged as part of the three-day festival. Top prize in 1948 went to Vantress Hatchery in California for its entry of a Cornish crossed with a New Hampshire; this company quickly became a recognized leader in the breeding of male broiler chicks and was later sold to Tyson Foods.

The global broiler chicken of today is a combination of Cornish male and White Plymouth Rock female lines.[17] Such breeds, referred to as 'industrial poultry stocks', are given numbers, not names, in order to control commercial information and dispersion. White birds are favoured because pinfeathers missed during plucking are less likely to be noticed by consumers. In 1920 broiler chickens averaged 1 kg (2.2 lb) when killed at around sixteen weeks of age; by 1941 they weighed 1.3 kg (2.9 lb) when slaughtered at twelve weeks; today, chicks average 2.7 kg (5.9 lb) when killed at six weeks of age.[18] Daily rates of growth have increased by 300 per cent (from 25 g or 0.88 oz to 100 g or 3.52 oz) over 50 years.[19] Moreover, the amount of food required to bring a broiler chick to slaughter weight today is radically less: 70 years ago 6.5 lb (3 kg) of feed was required to produce 1 lb (0.5 kg) of chicken meat; now it takes a mere 1.75 lb (0.8 kg).[20] 'The geneticists have thus created a bird that seems to confound common sense: it eats less and fattens faster.'[21]

Along with new scientific technologies, advances in selective breeding and the move to intensive confinement, another development changed poultry farming irrevocably. The entrepreneur Jessie Dixon Jewell is largely responsible for initiating the business structure known as vertical integration, according to which all aspects of chicken farming except the actual growing of the birds are brought under one giant operation. In the late 1930s Jewell, a transporter of chickens and eggs in Gainesville, northeast Georgia (whose town monument still boasts that the area is 'The Poultry Capital of the World'), organized local farmers to raise chicks for him if he supplied the birds and their food. After the Second World War this system, wherein larger corporations took charge of breeding, hatching, feeding, slaughter and distribution while farmers in effect leased birds and 'grew them out', proved extremely profitable because it lowered the price of chicken meat and so increased its popularity.[22]

Two of the major integrators from this period remain dominant today. The first, Delmarva-based Perdue Farms, invented a specialized feed from marigold blossoms, giving their chickens yellowish skin and richer meat. A one-time chief executive officer,

Broiler (meat) chick.

Frank Perdue, famously created the first recognizable poultry 'brand', Perdue Chicken, when in a 1970s TV commercial he declared: 'It takes a tough man to make a tender chicken' – at the same time marking the complete transition of chicken farming from a feminized to a masculinized endeavour.[23] The other chicken meat giant, Tyson Foods, was founded in 1935, and by 1990 was confident enough of its market share to use the ironic slogan: 'We're chicken'. Today, Tyson is the world's largest processor and marketer of chicken meat, working with 7,000 contract growers in the USA, processing 42.5 million birds a week, and supplying nuggets, wings and tenders to KFC, McDonald's, Burger King, Wendy's and Wal-Mart.[24]

By the 1980s, 95 per cent of broilers in the United States were from vertically integrated operations. The 'broiler belt', comprising Arkansas, Georgia, Alabama and Delmarva, was swamped with broiler sheds. Concerns were being raised about the conditions for slaughterhouse workers and the environmental impact of waste, including dead birds and manure. The industry responded by expanding into Kentucky, where the population of broiler chicks increased from 1.5 million to 231 million between 1990 and 2001.[25]

Kentucky, of course, was already synonymous with chicken meat, due to another crucial player. Legend has it that, in 1952, at the age of 65, Harland D. Sanders started his popular fast-food restaurant chain with a $105 Social Security cheque. Previously, while pumping petrol and cooking for weary travellers at a gas station in Corbin, Kentucky, Sanders developed the celebrated 'secret recipe' of eleven herbs and spices that would propel him to fast-food stardom. Becoming a local hero, Sanders was made a colonel, 'a political favour', not a military honour, by the Governor of Kentucky.[26] He travelled across America selling the Kentucky Fried Chicken franchise, receiving a nickel for every chicken sold. In 1964 Sanders traded in his interest in the company but remained its highly visible frontperson. In 1986 the franchise was taken over by PepsiCo, and five years later changed its name to KFC: its online history boasts that today 8 million customers are served each day in more than 11,000 restaurants across more than 80 countries.[27]

The history of 'fast chicken nation' does not stop with the colonel and his franchise. Fierce competition arose in the early 1980s when McDonald's introduced its patented 'McNuggets', derived from Tyson broiler chicks, notorious for their massively inflated breast meat, but consisting of 'small pieces of reconstituted chicken . . . held together by stabilizers, breaded, fried, frozen, then reheated'.[28] The advent and popularity of fast-food chicken profoundly affected the systems for raising and processing chicken meat around the world, while dramatically changing the eating habits and health of Americans – for example, one piece of KFC original chicken contains 1,315 kilojoules or 314 calories, which takes a 5 km run to burn off.[29]

While America continues to idolize fried chicken, the greatest activity in broiler production currently occurs in Asia and South America. In China in the early 1980s chickens were typically kept in small flocks in backyards; by 1997 more than 63,000 concentrated animal feeding operations (known as CAFOS) had emerged there, with single farms housing up to 10 million birds.[30] Indeed, consumption of chicken meat in China is estimated to have risen by 50 per cent over the last decade (substantial increases in broiler meat consumption have also occurred in Brazil, Mexico, South Korea, Taiwan, Indonesia and the Philippines).[31] But while the image of Harland Sanders remains synonymous with KFC in Western countries, children in Beijing, where this franchise was first introduced to China, apparently did not care for the colonel, whom they identified as a strict grandfather. Instead, the mascot for KFC China became the fun-loving 'Chicky', a cartoon-like white-feathered chicken dressed in red, white and blue, and wearing a baseball cap tilted sideways like a hip-hop artist.[32]

MEAT BABIES

Broiler chicks are conceived at specialized breeding facilities. Hens 'parent' chicks only until laying. Fertilized eggs are incubated mechanically and hatched chicks are transported to windowless broiler sheds or 'grow-out' houses, typically 400–500 by 40–46 feet (120–150 by 12–14 metres) where they live for six weeks in crowds of 10,000 to 30,000. Females may be killed as early as three weeks as 'Cornish game hens', while males may live three months if raised to be 'roasters'. Typically, broiler chicks are exposed to 23 hours of dim lighting for every one hour of darkness, in order to minimize activity while increasing appetite, and they stand or lie on litter that remains unchanged for the

Life inside a modern-day broiler shed.

duration of their lives. At slaughter age they are still juveniles, with the soft feathers and chirp-like vocalizations of chicks.

The European Union Science Committee stipulates a stocking density for meat chicks of no more than twelve birds per square metre, but in practice they tend to be housed more 'economically'. The UK, where 98 per cent of chicken meat production is intensive, allows up to nineteen birds per square metre; New Zealand and Australia permit twenty. On average, each broiler chick has a personal space smaller than an A4 sheet of paper. The mortality rate prior to slaughter is around 5 per cent in Britain, which means 45 million chicks die annually in 'grow out' houses before reaching 'market weight'.[33]

Inside the sheds, the noise of thousands of birds and the stench of their waste are overwhelming. Farmers, however, seldom endure these conditions, except during brief checks for faulty equipment or for dead or sick birds, since most chores associated with egg or chicken meat production are automated. In 1940, 250 'man-hours' were needed to 'grow' a thousand birds to 'market weight'; by 1955 the time required was around 48 hours.[34] As one industry promotional blurb puts it, 'When you choose a career in the poultry industry you may not see a chicken or an egg or a turkey – except at mealtime.'[35]

Broiler chicks have been selectively bred to grow fatter faster, which places intense pressure on their bodies: muscles and fat outgrow skeletons. Like other birds, chickens have delicate bones adapted to flight and quick movement, but the bones of broiler chicks are loaded with massively disproportionate breast weights. Consequently, their legs are often twisted and malformed. Up to 30 per cent suffer severe lameness and swelling, and at least that many suffer chronic pain. Veterinary scientists at Bristol University have shown that broiler chicks will self-medicate with food containing an unpleasant flavoured painkiller called Carprofen, and that the amount of the drug ingested increases with the severity of lameness.[36] This clearly demonstrates that broiler chicks routinely suffer pain and seek relief from it, despite the bitter taste of the analgesic. Currently, pain relief is not administered to broiler chicks or battery hens because chicken meat sold for consumption must derive from birds that have been free of drugs for at least 28 days. Unfortunately, this is the exact period when leg injuries are most prevalent.

By the time that their six weeks are up, broiler chicks find it hard to walk at all, and most spend 90 per cent of the time lying down in soiled litter. Many collapse for good: farmers refer to this condition as 'off their legs'. Being stranded on their own waste causes breast blisters and foot-pad dermatitis, and hock burns in chicks, effects sometimes observable in the bruising

seen on their carcasses in the supermarket. Broiler chicks also succumb to heart failure, liver disease and fluid build-up in the abdomen due to organ system failures. The larger males are particularly prone to Flip-Over Syndrome: sudden death by heart failure preceded by frantic wing flapping, convulsions and collapse.

The emotional lives of which chickens are capable (described in chapter Two) are for broiler chicks just as distorted and restricted as their physical forms. They are so overwrought by the compromised state of their bodies and their overcrowded, relentlessly noisy and filthy environment that any unexpected sound or movement can cause a mass panic, a phenomenon termed 'broiler hysteria' within the industry.

Few consumers are aware of the existence of specialized 'breeders' (to use the agribusiness term for them): these are chickens kept in flocks of several thousand with a male-to-female ratio of around one to fifteen. There are currently up to 56 million breeder chickens in the United States, and their lives are, if anything, more nightmarish than those of broiler chicks. Since meat chicks arrive at their massive 'market weight' and are slaughtered before they reach sexual maturity, those allowed to reach adulthood are even more enormous. For their first six weeks they are fed maximum amounts to determine how fast and big their offspring are likely to grow in the same timeframe; then their abundant food supply is suddenly cut and they are kept alive to mate on 80 per cent less food than chickens eat in nature.[37] Desperately hungry, breeder cockerels attempt to obtain the food of others: to prevent this, the nasal openings in their beaks are pierced with horizontal sticks so that their heads cannot pass through bars between cages. Males have spurs cut off, combs seared (in a procedure called 'dubbing') and beaks removed, because such giant roosters easily harm hens during

Due to their genetic propensity to 'bulk up', chicks raised for meat are crippled within a few weeks of life. These birds, rescued by Chocowinity Chicken Sanctuary, did not survive long after this photo was taken.

copulation. Research shows that 50 per cent of broiler chicken matings are forced, and hens suffer lacerations to heads, torsos and wings.[38]

'NATURAL' LAYERS?

In 2007, in the United States, more than 77.3 billion eggs were produced by 280 million hens (an average of 275 eggs per hen).[39] The lives of battery hens – intensively farmed egg-layers – are no less unhappy than those of broilers, but are much more visible to consumers due to decades of activism and open rescues. Although layer hens will not be baby chicks when dispatched to the slaughterhouse, they will have endured up to two years incarceration in tiny cages, subject to persistent noise, toxic smells and almost constant lighting. Seventy years ago a battery farm housing 100,000 hens would have been considered immense; today, it is not unusual for egg farms to keep 10 million birds at a time.

Chicks in hatching trays. The first day of life for those born into commercial egg farming.

158

The expendable male chicks of the egg industry.

There are no cockerels here. Male chicks, extraneous to the egg industry (except as breeders), are destroyed within twenty-four hours of hatching. Each year in the US alone more than 272 million male chicks are disposed of by gassing, microwaving, smothering or maceration (also termed 'instantaneous fragmentation'), their collective remains used as pet food.[40] Industry experts claim that fragmentation by fast rotating knives is the most humane method of extermination because it is the quickest,[41] yet the process is seldom highlighted by egg producers since it seems unlikely to win favour with a public for whom baby chicks are synonymous with cuteness, Easter, springtime and new life.

Layer pullets (young female chickens) are reared on deep litter or in cages until they are transferred to battery farms at about four months, when they begin laying. The remainder of their lives is spent inside cramped cages along with between four and nine cage mates. Thousands of identical cages are lined up

Layer hens on a battery farm.

in rows and stacked in vertical tiers. Water is supplied through nipple drinkers and food from a trough through the cage wires in front. Excrement drops through the floor of the cage for collection later. Eggs roll along a gradient into collection troughs. The birds experience unnatural lighting for around seventeen hours of the day, although those on lower tiers live in constant gloom.[42]

The advantages of battery operations for farmers are that thousands of birds can be controlled efficiently and economically, while eggs are easier to collect and less likely to be damaged during the laying process. For the birds, however, the constrained conditions do not allow even minimal normal behaviours. The area required to preen, scratch or merely turn round in is about three times greater than the space provided. Sometimes the height of the cage is not even enough to permit hens to stand properly.

Studies have shown that battery hens, typically deprived of the opportunity to dust-bathe, will work intensely to get access to dust-bathing substrate and engage in this activity.[43] Other experiments have shown that hens are desperate to arrange a nest before laying; nest-building is associated with natural changes in hormone levels occurring around the time of egg release. Caged hens become restless, 'show[ing] stereotypic pacing and escape behavior, or [they] perform "vacuum" nesting activity, the expression of the motions of building a nest in the absence of appropriate nesting materials'.[44] They also bully each other out of frustration and boredom.

Like broiler chicks, battery hens are fed special diets to make them more productive. Medicinal supplements are added to food to prevent infections associated with overcrowding, as well as dyes such as canthaxanthin that create yolks warmer in colour than the pallid ones produced by stressed birds.[45] Because they lead enforced sedentary lives and have been selectively bred to produce up to 300 eggs per year, battery hens suffer from calcium deficiencies leading to weak bones and fractures (also called 'cage layer fatigue'). In Europe around a fifth of hens in battery flocks have broken legs by the time they are collected for the slaughterhouse. Caged hens also suffer from deformed feet and claws, fatty livers, mouth ulcers, bronchitis, egg peritonitis, ascites, lung and heart damage, and are more prone to developing tumours. One of the worst conditions affecting commercial layer hens, prolapsed uteri, results from the manipulation of breeds in order to induce early and more frequent egg production.[46] Prolapses are also caused by caged birds straining to lay eggs in situations far removed from their natural proclivities.

Supporters of contemporary poultry farming argue that aggressive behaviours such as feather and vent pecking that result from hostile living conditions can be controlled within intensive operations.[47] Management methods include 'beak trimming', a procedure whereby the tips of hens' upper beaks are seared off with hot blades and cauterized to halt bleeding. Beak searing is considered more economical if performed rapidly and without anaesthesia or pain relief. Yet beaks are sensitive organs comprising free nerve endings that act as sensory receptors, and beak-trimmed hens are often left with chronic pain, compromised eating ability and weakened senses of taste, smell, temperature and touch. Already illegal in Sweden, Norway and Finland, and to be banned in the UK from 2011, beak searing may eventually be abandoned in favour of the selective breeding of supposedly more passive layer hens.

Hens in a battery shed are usually killed at around eighteen months of age and replaced with a new flock of pullets. Some, however, are spared for another laying season; this requires forced molting, the practice of depriving hens for several days of food

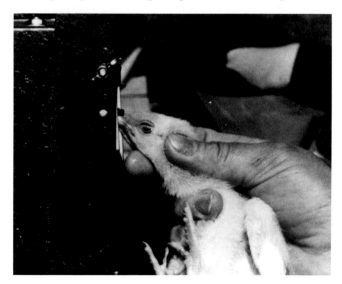

Searing the beak off a chick destined for a battery farm.

(and sometimes water) so the hormone changes responsible for both new plumage and the resumption of laying occur after only a few weeks, rather than four or more months (as would transpire naturally). Outlawed in both the UK and the USA since 2006, forced moulting nevertheless continues to be routinely practised by many egg producers.

Layer hens are also deprived of food for a few days prior to slaughter since savings incurred by starving 'end of lay' hens off-set the costs associated with their transportation to 'processing'. The bodies of 'spent' hens are too damaged from factory life to be displayed to consumers as meat. Instead, they are shredded to be made into soups, pies and other institutional foods, or turned into feed for farm animals, including other hens.

ALTERNATIVE FARMING SYSTEMS

The vast majority of chickens in agricultural settings are not provided with the five basic freedoms listed by the Farm Animal Welfare Council as indexes of good animal welfare.

Lack of space is one of the chief concerns of avian behaviourists regarding intensive farming of chickens for meat. In response to growing consumer discomfort around such issues, some farms permit chicks to venture outside. These so-called free-range systems, where birds are raised together in reduced numbers for up to eleven weeks and have the opportunity to experience sunlight and to perch, are arguably more humane. However, observation of free-range broiler chicks indicates that many do not roam beyond the inside of the shed. One reason is that without their mother hen, young chicks are fearful of new situations; another is that outside the shed they are not provided with the tree cover that chickens prefer in order to hide from predators.[48] The main problem, however, is the genetic make-up of these birds: leg weakness, fractures and chronic pain caused by genetically manipulated obesity mean that even the chicks who venture outside tend to lie down rather than engage in normal free-ranging activities.

In response to criticism of battery systems, the egg industry often claims that 'only happy hens lay eggs'. In fact, hens lay whether they are happy or not, because creating an egg is an inbuilt biological process, and one artificially exaggerated by the extended daylight hours and other methods of intensive egg farms.[49] Some countries such as Switzerland and Sweden, along with the state of California, have already outlawed battery cages for laying hens, and the European Union has pledged to phase them out by 2012. Meanwhile, although most hens in the egg industry remain in intensive operations, around 4 per cent in industrialized countries live in one of four recognized alternative systems where expression of natural behaviour is less restricted: these systems include enriched cages, aviaries, colony/barn or deep litter and free-range. An enriched cage provides a cramped flock with a perch, nest area and dust-bathing box. In aviary systems hens live in sheds stocked with levels of perches. Around 4,000 hens are held in each colony, which consists of a windowless barn in which birds may be able to perch and nest. Free-range hens have access to an outside environment and can, in theory, forage and dust-bathe within a restricted area. In practice, however, many hens in these systems remain inside, overwhelmed by the sheer number of flock mates encountered when trying to enter or leave sheds. Only smaller flocks really enable chickens to behave as they would in the wild, because they can recognize individual flock mates and maintain peck orders for safe social interactions. Birds raised for commercial purposes, because of the numbers involved, are 'in a constant state of trying to establish a hierarchy but never achieving it'.[50]

Hens on a free-range egg farm. Smaller flocks allow chickens to live more as they would in nature.

The final stage of industrial chicken farming, whether for meat or eggs, occurs at the slaughterhouse or 'processing plant'. In the USA more 'end of lay' hens are killed each year than pigs or cows.[51] The slaughterhouse has undergone dramatic changes as a result of technological developments since the Second World War. As mentioned, dismemberment of chickens was not routine during the early part of the twentieth century: evisceration was pioneered during the 1930s, but it was over the next two decades that the sale of ready-to-cook carcasses prospered. In the USA in 1965, 78 per cent of chicken carcasses were sold whole to consumers, 19 per cent were sold cut up and 3 per cent processed further; by 2005 only 11 per cent were sold whole, 43 per cent were cut up and 46 per cent went on to further processing.[52]

Nowadays, slaughterhouses are highly mechanized assembly lines, factories that deal with more than 8,000 chicken carcasses per hour.[53] On arrival at a plant, broiler chicks and 'spent' battery hens are hung upside down from metal shackles and sent along a moving line in a dimly lit room. The rationale for low illumination is economic: if chickens cannot easily see their flock mates being stunned they are less likely to struggle and incur an injury that would downgrade the carcass. The line first takes the hanging chickens to a shallow saline bath in which they are stunned when their dangling heads touch electrified water. At the next station their throats are mechanically cut by rotating blades and the birds bleed to death in 'the bleed tunnel'. The carcasses are then transported to the scalding bath where a series of 'rubber finger' machines de-feather each body before it travels through flames to burn off any residual signs of plumage. Feet (or 'paws' as they are referred to in the industry) are chopped off next, after which the carcasses tumble down a chute to workers who manually re-hang them by the wings before the evisceration process commences. Entrails are hooked out by piston-like plungers and inspected by government officials for signs of disease or damage. Gizzards, hearts and livers are separated, in some cases to be recombined and later inserted back into whole chickens; lungs are removed by a suction machine, and the carcasses come to rest in the chill bath for a few hours.

Following chilling, the chickens' bodies are dismembered in the cutting room. This part of the processing is relatively new, and caters for the rising demands for 'convenience' chicken meat items such as breasts, wings, thighs and drumsticks. It is performed by machines, but workers are required to position the carcasses. If portions are going to KFC joints, the eleven special herbs and spices are pinpricked under the skin prior to distribution. Packaging and delivery conclude the conversion of chickens from animate beings to edible objects. Broiler chicks end up on supermarket shelves. Legs and feet are sent to Asian countries where they are delicacies. As mentioned, 'retired' layer hens end up in pet food or are ground up and fed back to other hens.

At every stage of processing prior to slaughter, chickens are vulnerable to extreme suffering. Catching involves teams of workers clearing broiler or layer sheds of all their birds at night. Since profits rely on speed, chickens cannot be handled individually, but are carried to waiting crates, sometimes up to six per hand, upside down by the legs or wings. As a result, leg fractures, broken or dislocated wings and feet and ripped claws are common.[54] Some farms use automatic 'chicken harvesters' that 'hoover the birds up with a giant nozzle and take them into the machine's belly', since these are said to lessen the risk of physical injury to frightened birds and thereby preserve the quality of carcasses.[55]

Transportation to the slaughterhouse involves up to 6,000 terrified chickens crammed together in hot and dirty conditions; many will be thirsty (they are given no food or water prior to dispatch) and always some will die from stress, injury in transit, or over-heating (chickens are not easily able to regulate body temperature). Many also suffer fractures or dislocations when rapidly hung on the assembly line. As for the supposedly humane step of stunning before slaughter: statistics from the US Department of Agriculture's Food Safety and Inspection Service reveal that in the USA, where the well-being of chickens is not covered by the federal Animal Welfare Act or the Humane Slaughter Act (despite birds comprising 95 per cent of the non-aquatic animals killed for food each year), around 2.8 million broiler chicks were boiled alive in de-feathering tanks in 2002, having passed above the stunning baths and missed the blades at throat-cutting stations.[56] Boiling alive tends to occur to smaller chicks whose heads

Loading chickens onto the slaughter line.

do not reach stun baths, and to battery hens, who remain more alert and active than their already crippled broiler cousins.

In the UK, where up to 50 birds an hour are conscious when their throats are cut, and up to 9 in 1,000 birds survive the blade and perish in scalding tanks, the Humane Slaughter Association has argued that chickens should be killed or rendered completely unconscious via gassing or stunning.[57] The poultry industry, however, resists adopting electrocution as a method of killing because 'over-stunning' results in bruises and fragments of shattered bone appearing in meat. Consumers' preference for the appearance of pure white meat shapes the decisions made about how birds are slaughtered.

The front-line processing of chicken meat often involves human suffering as well. In the USA, workers in chicken meat plants tend to be unskilled peoples from underprivileged populations.[58] They are paid low wages, especially if working in the non-unionized states in the Southern 'broiler belt', where chicken farming is most common. Slaughterhouses are noisy, hot and smelly. The air is constantly full of feathers, faeces, the sound of terrified birds and loud machinery. The work is relentless and exhausting: the lines move continuously so employees have few breaks, and the rapidity of the process causes repetitive strain injuries and accidents. The monotony and speed increase desensitization to the chickens' suffering; abuse of birds is routine, and includes the stomping, crushing, punching, kicking, spray painting and 'throwing for fun' of live chickens.[59]

POLLUTION AND DISEASE

The trade association Delmarva Poultry Industry Incorporated declares on its news and information webpage that 'every day is Earth Day for America's chicken industry' – a perplexing statement from a business that creates vast water, air and land pollution.[60] The same webpage endorses the policy of recycling litter in chicken houses between flocks, 'meaning less has to be removed and land-applied', a seemingly eco-friendly solution that nevertheless causes great distress to the chickens. Another way of 'recycling' poultry litter in the USA is to feed it to cows – as much as three tons of poultry waste can be consumed by a single cow in a year.[61]

Intensive farming of chickens contributes to water depletion, land degradation and deforestation because of the large quantities of space and water required to grow feed for confined chickens.[62] Meanwhile, chicken slaughterhouses typically use more than 2 million gallons of water per day.[63] In the Delmarva Peninsula alone, the raising and killing of 600 million chickens per year results in 3.2 billion pounds of raw waste, 13.8 million pounds of phosphorus and 48.2 million pounds of nitrogen.[64] A million-hen complex produces 125 tons a day of wet manure – a virulent pollutant if it enters water supplies.[65] In 1998 Tyson Foods was fined $6 million for polluting a waterway in Maryland with coliform bacteria, phosphorus and nitrogen from one of its slaughterhouses.[66] Airborne contaminants such as feather barbules, skin and ammonia from faeces produce health problems such as asthma, headaches, infections and mood disorders for those working on farms and in slaughterhouses, or living near plants. The broiler industry produces up to 3 billion pounds of feathers per year. In Europe these are land-filled while in the United States they are autoclaved to produce material that goes into feed for farmed animals and pets.[67]

As well as waste disposal problems, intensive farming breeds and spreads disease. The close packing of birds means that bacterial and viral infections travel rapidly through flocks. Although confined birds 'may be free from the parasitic infestations the early battery farmers sought to overcome, "diseases of intensification" have taken their place'.[68] The H5N1 virus was probably caused, and certainly exacerbated, by the unnatural compaction of thousands of chickens. All flu viruses affecting birds start out low grade and are harmless to humans. Some mutate into more virulent 'fowl-plague' forms, but this requires a virus present in a wild bird in its mild form to be introduced to poultry – something made much more likely when chickens' immune and respiratory systems have been compromised by stress, filth and close confinement. In contrast, outdoor flocks are additionally protected because sunlight kills such viruses.[69] The poultry industry moves in the face of impending flu disasters by reassuring the public that avian viruses, transmittable to other species, can be contained by killing all birds in the vicinity of infected areas and then by more intensely confining farmed chickens – the same solution that enabled the problem to arise in the first place.

Chickens suspected of harbouring the 'bird flu' virus are collected for mass killing, 2000.

The early twentieth century saw the transformation of chicken meat from an expensive and scarce delicacy into a widely available and affordable commodity; the 1980s brought condemnation of 'fatty' red meats and promotion of white meat's nutritional benefits; the 1990s endorsed the convenience of frozen chicken meals and fast foods.[70] The first decade of the twenty-first century can be characterized by increasing public awareness of where foods come from as well as the risks associated with certain forms of production and consumption. The high-volume processing that has made chicken meat 'an inexpensive and popular meal' has also led to hazards of contamination and infection.[71] One USDA microbiologist argues: 'there are about 50 points during processing where cross-contamination can occur. At the end of the line, the birds are no cleaner than if they had been dipped in a toilet.'[72] Examples include the de-feathering and chilling processes, which create dust that transports bacteria readily, or the scalding process, during which the excrement from terrified birds builds up to create what slaughterhouse workers refer to as 'faecal soup' in the tanks.

Over the past century, in the push to subordinate the natural proclivities of chickens to the 'dictates of industrial production', these birds – once admired and respected – have been deprived of their 'own biological foundation'.[73] They have lost their natural births and relationships with mother hens and siblings to incubators and brooders; their immune systems to compulsory antibiotics; their connections to the earth and sky to confinement in overcrowded barns; their enjoyment of nesting and dustbathing to barren cages; their sociability to the constant stress of immersion in a crowd of strangers; their love of sunlight and the changing seasons to Vitamin D regimes and artificial lighting; their capacity to fly, run or just stand to enforced obesity. The hen's delight in springtime broodiness has been subjugated to unseasonal hyper-production of ova. No longer a mother, she is a 'breeder' or an 'egg-laying machine'. The newly hatched male chick of the egg industry moves straight down a chute to maceration. And to those that survive their time in the factory farm one final indignity is granted: an assembly-line death that may be prolonged, painful and terrifying.

When the genome of the red jungle fowl was sequenced in 2004, one of the researchers involved enthusiastically declared that this new knowledge about chickens would act as a '"bible" for those who seek to breed [even] faster-growing birds, lower-fat breasts and more prolific egg-layers'.[74] Two years previously a featherless chicken was 'developed' by geneticists at the Hebrew University of Jerusalem. Described as 'lower in calories' and

The new breed of genetically modified featherless chicken developed by Avigdor Cahaner at the Hebrew University of Jerusalem, Rehovot, Israel.

'environmentally friendly' (no plucking required), this chicken is likely to revolutionize mass poultry farming in warmer countries. The invention of ready-made naked meat chicks (able to be packaged within a month of hatching) and 'egg machines' capable of laying 365 days of the year is imminent. Global chicken farming seems set to intensify even further.

Epilogue: Appreciating Chickens

The past century has certainly brought profound changes to the quality of life for the vast majority of the planet's chickens. However, the short and miserable lives of intensively farmed birds have not gone unnoticed or unchallenged; nor has industrialization been successful in eradicating more sanguine connections to these birds from our everyday lives. In fact, alongside the developments in poultry farming discussed in the previous chapter, there has been a significant resurgence of interest in chickens as backyard helpers and cherished companions. The growing appreciation of chickens as pets occurs as part of a wider re-evaluation of – and nostalgia for – the natural world and a desire to engage more closely with nonhuman species. Evident since the 1970s, this trend is associated with an increase in recreational activities that involve animals, a dramatic rise in the number of households containing companion animals, enhanced public interest in animal welfare issues, and higher visibility of animal advocacy groups.[1] And, like other animals, chickens have benefited from these recent socio-cultural and attitudinal changes.

A rehomed ex-battery farm hen, now enjoying her new life outdoors.

The rise of the urban chicken movement has been noticeable over the past decade or so. Fostered at first in London following the invention there of the 'Eglu' – a specialized henhouse for city chickens – urban flocks quickly took off in North America and Australasia as well. The recent flurry of books endorsing keeping chickens for their fresh eggs and idiosyncratic company is testament to the popularity of this movement. Not all cities, however, permit 'livestock' to be kept in backyards. In Madison, Wisconsin, illegal chicken-keepers, known as 'the poultry under-ground', decided to campaign in order to change local laws that forbade chickens within the city limits. Granted the right to raise hens (not roosters) in backyards from 2004, this group immedi-ately set up 'Chicken 101' classes and hosted neighbourhood 'Tour Des Coups', for which members of the public purchased maps in order to visit chickens 'at home' and learn about responsible chicken guardianship.[2]

Positive attitudes towards chickens have also been fostered through recent documentaries profiling real-life relationships people experience with *Gallus domesticus*. *The Natural History of the Chicken*, released in 2000, bridges the terrain between com-panionship with and advocacy for chickens, tracing the many faces of human–chicken interactions – from the pampering of pet roosters, to the breeding of gamecocks for fighting, to the raising of battery hens and broiler chickens.[3] In the director Mark Lewis's trademark quirky fashion, the chickens in this award-winning film find themselves in unusual situations (wearing diapers, floating in a swimming pool or thawing out after human mouth-to-chicken mouth resuscitation), but it is their human companions who appear most eccentric. Yet these moments come across as understandable and humane – all

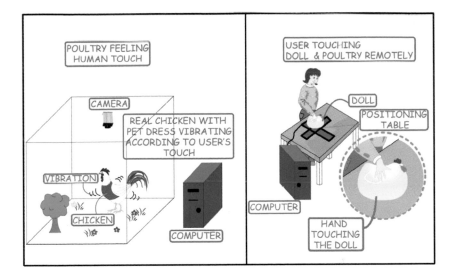

the more so when juxtaposed with footage from battery farms that reveal the constant level of noise, light and chaos for caged hens.

The role of chickens as companion species is also being regarded seriously in the Mixed Reality Lab at the National University of Singapore, where a revolutionary means for humans and chickens to maintain tactile proximity while separated geographically has been developed. Motivated to alleviate the isolation experienced by chickens left alone when their human companions are at work, the researchers Adrian Cheok and James Teh employ Internet technology to generate human-to-pet interaction. The chicken wears a wireless, sensor rigged jacket called a 'hug suit' that simulates the sense of touch, while the human (in another location) has a doll, which, when touched, transmits the caressing sensations through the hug suit to the chicken. At the same time, the movements of the chicken are

Helping chickens and humans keep in touch (Poultry Internet schematic). Poultry Internet is the invention of James Teh and Adrian David Cheok at the Mixed Reality Lab, National University of Singapore.

Clare Druce (far right) and Violet Spalding (white hat) at an early Chickens' Lib protest outside MAFF, London.

watched on camera (not unlike the popular chicken-cam operations proliferating online).[4]

ADVOCACY

While enthusiasm for chickens as companions is returning, advocacy on behalf of these birds is also gaining more prominence. Campaigns against the mistreatment of chickens began in earnest in the 1970s when a small group of women protested on the steps of the Ministry of Agriculture, Fisheries and Food at Whitehall Place in London. A few days earlier, one of these women had talked her way into obtaining a couple of 'spent' battery hens destined for slaughter in the East End. These hens were taken to the protest in an imitation battery cage built by the woman's husband, providing visual testament to the condition of birds farmed intensively for their eggs throughout the UK. This action launched the public face of an organization called Chickens' Lib.

Led for many years by Clare Druce and her mother Violet Spalding, Chickens' Lib was the first group to dedicate its activities primarily towards raising awareness of intensive chicken farming. Starting out with just a handful of supporters, Druce's movement grew to have a huge impact on the British scene and also inspired chicken advocacy efforts across the Atlantic.[5] It did so by making effective use of media publicity and, most importantly, of visual representation.

The use of imagery remains crucial in contemporary animal advocacy, and works particularly well for issues relating to creatures that are considered charismatic (such as great apes and cetaceans) or cute (like kittens and puppies). The impact of imagery is more complex when publicizing the exploitation of chickens, since these birds do not possess many of the features

Protestors from the activist group 'Kentucky Fried Cruelty'.

likely to elicit a 'cute response', and are not readily perceived as special by a general population largely conditioned to view chickens as food even before their deaths. Despite these disadvantages, graphic campaigns against the battery farming of layer hens have proven effective, since only one look is really required to see what is objectionable about this. Efforts to raise awareness of egg farming have therefore relied heavily on the visual portrayal of real chickens and on spectacular footage of what goes on in sheds behind the scenes. Such raw footage may be obtained during 'open rescues', overt operations involving the release and rehabilitation of incarcerated hens.[6]

Even without explicit cues, consumers can be asked to visualize the invisible by imagining what life might be like for factory-farmed birds. The Pulitzer Prize-winning author Alice Walker used this tactic in a letter she wrote to the chief executive officer of the parent company of KFC, imploring him to imagine being a broiler chick.[7] This strategy is important for raising public awareness about the lives of meat chicks, since, unlike battery hens, these birds 'appear' to be healthier because they retain their feathers, are plump, seem greedy, and are not confined in cages.[8]

SANCTUARY

In many ways, the fashionable urban chicken movement, discussed above, has helped to educate the public and encourage more respectful understanding of chickens, but it also has its critics. In 2009 the downside of the waning enthusiasm of hobbyists for their flocks was evident in a dramatic rise in reported intakes of unwanted chickens – especially roosters – by sanctuaries and shelters. And, paradoxically, those seeking to set up new city flocks tend to obtain chicks from industry hatcheries, cleverly capitalizing on this trend, instead of adopting unwanted chickens from shelters. Chicken sanctuaries therefore take in the casualties of the urban chicken movement, in addition to rehoming rescued battery hens and broiler chicks, and any other outcast, injured or stray chickens.

Founded in 1990 by Karen Davis, United Poultry Concerns is the world's foremost non-profit organization dedicated to promoting the respectful treatment of domestic fowl. UPC runs a haven for chickens in Virginia, and also teaches people about the egg and chicken meat industries, the natural lives of free chickens, pleasures and benefits of human–chicken companionship, and alternatives to chicken farming and the use of chickens in education and scientific experimentation.[9] Other prominent sanctuaries in North America include Farm Sanctuary in upstate New York, Chocowinity Chicken Sanctuary in North Carolina, and Chicken Run Rescue, a rehabilitation and adoption centre operating from the basement of a home in Minneapolis. In the United Kingdom, The British Hen Welfare Trust finds refuge for thousands of commercial laying hens destined for slaughter, and also invites people to sponsor individual rescued hens.

Another sanctuary exists in Vermont catering specifically for the rehabilitation of former fighting roosters, birds often assumed to be hopeless cases due to incorrigible aggressiveness. The clinical psychologist pattrice jones, an expert on trauma recovery, and co-founder (with Miriam Jones) of Eastern Shore Chicken Sanctuary and Education Center, believes that these 'icons of courage' are in fact deeply 'frightened birds' whose early separation from hen and siblings, compounded by years of isolation tethered to outdoor stakes and repeated injections of drugs to heighten the natural flight or fight response, have prevented them from learning 'normal' gallinaceous social skills. Just as chickens have been shown to suffer from post-traumatic stress disorder, these birds may also be helped by the same kinds of therapeutic interventions used to treat anxiety and fear in humans, adapted within a model of 'trans-species psychology'. The specialized rooster rehabilitation programme at Eastern Shore Sanctuary consists of instilling a sense of safety in new arrivals (by first protecting them from all contact with other roosters), gradually and gently reintroducing the birds to the other chickens on the property, assisting the acquisition of less fear-driven and more acceptable modes of interacting with flock members, and encouraging 'time out' for aggressive behaviours.

Felipe is one of the success stories of this programme. He was confiscated from an illegal cockfighting operation, and his conditioning to automatic violence against other roosters made adjustment to life within a flock a particular challenge. Now he lives peaceably among the other chickens, preferring to do his own thing during the day, but each night seeking out and sleeping beside Fabio, another solitary rooster. Eastern Shore Chicken Sanctuary works according to the principle 'let birds be birds'; this means permitting resident chickens to have the greatest freedom possible to express their natural inclinations. Consequently, ex-fighting roosters mix and mate with rescued broiler and battery hens, producing mixed-breed offspring of their self-selected couplings. The birds 'rewild' themselves, forming feral flocks that roam the woods of the rural neighbourhood.[10]

THERAPY

It is now well recognized that sharing time with other species provides physical and emotional health benefits for humans. Animal assisted therapy (AAT) is the term given to activities involving particular animals as part of a person's or a group's therapy, rehabilitation or education. AAT has been used in rest homes, hospitals, schools and prisons for more than 40 years

Mr Joy with friend Kathryn Black. Like other animals, 'therapy chickens' provide uplifting company for those in rest homes and hospitals.

and has shown favourable outcomes for those with developmental, emotional, behavioural and physical health issues.[11] It is perhaps more commonly associated with dogs, cats or horses, but 'therapy birds' are also popular with groups such as children and the elderly (and sometimes the benefits of AAT can work both ways: in New Zealand, for instance, prisoners gain a sense of empathy for those less fortunate when they are given the responsibility to care for and rehabilitate former battery hens).

One of the most famous chicken therapists was an Old English Game rooster named Mr Henry Joy, who lived with Alisha

An audience with Mr Joy.

Tomlinson in North Carolina. When Mr Joy came to Tomlinson's home following the hospitalization of his previous human companion, he was suffering from a progressive foot disorder that led to gangrene and the amputation of most of his toes. It was while he was recuperating – travelling everywhere with Tomlinson, observing life from the safety of a basket – that his charismatic presence became apparent: Mr Joy cheered people up wherever he went. Soon the duo were visiting nursing homes and assisted living centres where the bantam imparted 'chicken therapy' to the elderly and infirm, his charity work gaining attention across the country. Tomlinson and Mr Joy also educated the public during media and community appearances (and via his website) about chicken sentience and intelligence, factory farming and the truth behind nuggets. Mr Joy died suddenly at the beginning of 2009, leaving behind two grieving hens and Alisha

Tomlinson, who is determined to continue his advocacy on behalf of chicken-kind: 'In his nine years of life, Mr Joy touched many lives proving that when it comes to love, neither size nor species are important . . . His activities changed the way most people relate to chickens as merely sandwiches, strips and buffalo wings.'[12]

Those campaigning for better lives for the chickens of this planet work hard to make their stories more visible and their pictures worth a thousand words. They also remind us that, although we have lived with chickens for thousands of years – although we have used them to make everything from myths to movies to McNuggets – we still have a great deal to learn from them. Maybe we would all benefit from an audience with Mr Joy.

Timeline of the Chicken

Jurassic period	Ecocene period	c. 8000–6000 BC	c. 1500–700 BC	Classical antiquity
Era of archae-opteryx, an early form of bird	The 'pre-chicken', direct ancestor of the jungle fowl, is thought to have evolved	Domestication of chickens started. Earliest remains, found in China and Pakistan, are at least 7,500 years old	Chickens emerge in hieroglyphics; ancient Egyptians use mass incubators to hatch eggs	Chickens used for food and cockfighting. Roosters appear on coins and artefacts as symbols of virility

late 1800s–early 1900s	1874	1879	1887	1902
'Hen craze' era in Britain and America. Chicken breeding popular. First poultry shows occur	Composer Modest Mussorgsky publishes *The Ballet of Unhatched Chicks* and *Hut on Chicken Legs*	The incubator is invented by Lyman Byce in Petaluma	Birth of Marc Chagall, an artist whose works often feature roosters	A town in Alaska calls itself 'Chicken'

1952	1973	1980s	1990
Harland D. Sanders launches his first fast-food restaurant chain, which would become known as Kentucky Fried Chicken, and later KFC	Clare Druce and Violet Spalding establish 'Chickens' Lib' in the UK, the first activist organization against the battery farming of chickens	McDonald's patents the 'McNugget' in a competitive move against Kentucky Fried Chicken	United Poultry Concerns is established by Karen Davis. Now the world's largest chicken advocacy agency

late 14th century AD	15th–17th centuries	1600	1789–99	mid-1800s
Chaucer's Chauntecleer appears as part of *The Canterbury Tales*	Chickens popular in Dutch art. Melchior d'Hondecoeter paints *A Cock, Hens and Chicks* around 1668	Italian naturalist Ulisse Aldrovandi publishes his famous treatise on chickens	The rooster becomes a recognized symbol of French nationalism around the time of the French Revolution	Cockfighting outlawed as a recreational 'sport' in Britain and the United States

1920s	1923	1928	late 1930s	1944
Petaluma known as the 'Egg Basket' of America. Battery farming increasingly popular	Wilmer Steele mistakenly receives 500 instead of 50 chicks from a hatchery. Year-round production of broiler chickens begins on Delmarva Peninsula	Presidential hopeful Herbert Hoover promises Americans 'a chicken in every pot'	Jesse Dixon Jewell initiates the business structure known as vertical integration, where farmers lease broiler chicks from corporations	Pablo Picasso paints *Cock of the Liberation* while living in Nazi-occupied France

2000	2002–3	2004	2007	2009
Dreamworks releases Claymation film *Chicken Run*. Mark Lewis directs *The Natural History of the Chicken*	Fear about new outbreak of 'bird flu'. Featherless chicken created	Chicken's genetic code is cracked. First avian genome to be assembled – poultry industry celebrates	*Attention Chicken!* Gigantic meat chick found in supermarket car park in Milwaukee	Mr Henry Joy dies: remarkable rooster, activist and chicken therapist

References

1 FROM *T. REX* TO TRANSYLVANIAN NAKED NECKS

1 J. Long and P. Shouten, *Feathered Dinosaurs* (Oxford, 2008), p. 3.
2 J. Asara et al., 'Protein Sequences from Mastodon and
   Tyrannosaurus rex Revealed by Mass Spectrometry', *Science*,
   cccxvi/5822 (2007), pp. 280–85.
3 R. D. Crawford, *Poultry Breeding and Genetics* (Amsterdam, 1990),
   p. 4.
4 J. del Hoyo et al., *Handbook of the Birds of the World*, vol. ii
   (Barcelona, 1994), pp. 529–31.
5 David M. Sherman, *Tending Animals in the Global Village*
   (Philadelphia, 2002), p. 47.
6 M. C. Appleby, J. A. Mench and B. Hughes, *Poultry Behaviour and
   Welfare* (Cambridge, 2004), p. 3.
7 A. Fumihito et al., 'Monophyletic Origin and Unique Dispersal
   Patterns of Domestic Fowls', *Proceedings of the National Academy
   of Sciences of the USA*, xciii (1996), p. 6792.
8 J. Eriksson et al., 'Identification of the Yellow Skin Gene Reveals a
   Hybrid Origin of the Domestic Chicken', *Public Library of Science
   Genetics*, iv/2 (2008). See www.plosgenetics.org/article/info:doi/
   10.1371/journal.pgen.1000010
9 Appleby, Mench and Hughes, *Poultry Behaviour and Welfare*, p. 3.
10 Page Smith and Charles Daniel, *The Chicken Book* (Athens, GA,
   2000), p. 14.
11 Roger Blench and Kevin C. MacDonald, 'Chickens', in *The
   Cambridge World History of Food*, ed. K. F. Kiple and K. C. Ornelas
   (Cambridge, 2000), ii.g.6, vol. i.

12 Crawford, *Poultry Breeding and Genetics*, p. 14.

13 Smith and Daniel, *The Chicken Book*, pp. 28–31.

14 Appleby, Mench and Hughes, *Poultry Behaviour and Welfare*, p. 2.

15 Crawford, *Poultry Breeding and Genetics*, p. 43.

16 Ibid., p. 44.

17 Cited in Pam Percy, *The Field Guide to Chickens* (Stillwater, MN, 2006), p. 25.

18 Carol Ekarius, *Storey's Illustrated Guide to Poultry Breeds* (North Adams, MA, 2007), p. 6.

19 Chris Graham, *Choosing and Keeping Chickens* (London, 2006), p. 73.

20 Ibid., pp. 92–4.

21 Percy, *Field Guide to Chickens*, p. 27.

22 Graham, *Choosing and Keeping Chickens*, p. 80.

23 Ekarius, *Storey's Illustrated Guide*, p. 131.

24 Ibid., p. 83. The American Livestock Breeds Conservancy (ALBC) defines 'critical' as there being fewer than 500 breeding birds in the United States, with five or fewer primary breeding flocks (50 birds or more), and globally endangered.

25 Ibid., p. 103.

26 Ibid., p. 80.

27 Graham, *Choosing and Keeping Chickens*, p. 176.

28 Ekarius, *Storey's Illustrated Guide*, p. 160.

29 Pam Percy, *The Complete Chicken* (Stillwater, MN, 2002), p. 59.

30 Graham, *Choosing and Keeping Chickens*, p. 130.

31 Ibid., p. 144.

2 CHICKEN WISDOM

1 T. J. Eddy et al., 'Attribution of Cognitive States to Animals', *Journal of Social Issues*, XL (1993), pp. 87–101.

2 T. X. Barber, *The Human Nature of Birds* (New York, 1993), p. 109.

3 S. E. Orosz and G. A. Bradshaw, 'Avian Neuroanatomy Revisited', *Veterinary Clinics: Exotic Animals*, X (2007), p. 775.

4 Lesley Rogers, *The Development of Brain and Behaviour in the Chicken* (Wallingford, Oxfordshire, 1995), p. 231.

5 Ibid., p. 48.

6 Ibid., p. 53.

7 Ibid., p. 56.

8 Ibid., p. 84.

9 Ibid., p. 96.

10 Gordon Burghardt, *The Genesis of Animal Play* (Cambridge, MA, 2005), p. 273.

11 Marian Stamp Dawkins, 'What Are Birds Looking At? Head Movements and Eye Use in Chickens', *Animal Behaviour*, LXIII (2002), pp. 991–8.

12 Marian Stamp Dawkins, 'How Do Hens View Other Hens? The Use of Lateral and Binocular Visual Fields in Social Recognition', *Behaviour*, XII (1995), pp. 591–606.

13 M. Appleby, J. A. Mench and B. Hughes, *Poultry Behaviour and Welfare* (Cambridge, MA, 2004), p. 18.

14 Dawkins, 'What Are Birds Looking At?' p. 991.

15 Amy Hatkoff, *The Inner World of Farm Animals* (New York, 2009), p. 22.

16 Ibid., p. 33.

17 G. Vallortigara, 'The Cognitive Chicken: Higher Mental Processing in a Humble Brain'. See www.science.org.au/sats2007/vallortigara.htm

18 S. M. Abeyesinghe et al., 'Can Domestic Fowl, *Gallus Gallus Domesticus*, Show Self-Control?', *Animal Behaviour*, LXX (2005), pp. 1–11.

19 G. McBride et al., 'The Social Organization and Behaviour of the Feral Domestic Fowl', *Animal Behaviour Monographs*, II (1969), p. 25; D.G.M. Wood-Gush and I.J.H. Duncan, 'Some Behavioural Observations on Domestic Fowl in the Wild', *Applied Animal Ethology*, II (1976), pp. 255–60; N. E. Collias et al., 'Locality Fixation, Mobility and Social Organization within an Unconfined Population of Red Jungle Fowl', *Animal Behavior*, XIV (1966), pp. 550–59.

20 Wood-Gush and Duncan, 'Behavioural Observations on Domestic Fowl', p. 255.

21 M. Zuk et al., 'Male Courtship Displays, Ornaments and Female Mate Choice in Captive Red Jungle Fowl', *Behaviour*, XII (1995), pp. 821–36.

22 M. Leonard and L. Zanette, 'Female Mate Choice and Male Behaviour in Domestic Fowl', *Animal Behaviour*, LVI (1998), pp. 1099–1105.

23 D. R. Wilson et al., 'Alarm Calling Best Predicts Mating and Reproductive Success in Ornamented Male Fowl, *Gallus gallus*', *Animal Behaviour*, LXXVI (2008), pp. 543–54.

24 Cited in M. Hauser, *Wild Minds: What Animals Really Think* (London, 2001), p. 216.

25 C. S. Evans, 'Cracking the Code: Communication and Cognition in Birds', in *The Cognitive Animal*, ed. M. Bekoff, C. Allen and G. M. Burghart (Cambridge, MA, 2002), pp. 315–22.

26 M. Leonard and A. G. Horn, 'Crowing in Relation to Status in Roosters', *Animal Behaviour*, XLIX (1995), pp. 1283–90.

27 See http://www.time.com/time/magazine/printout/ 0,8816,875688,00.html

28 D. G. Griffin, *Animal Minds: Beyond Cognition to Consciousness* (Chicago, 2001), p. 172.

29 Ibid., p. 173.

30 Evans, 'Cracking the Code', pp. 315–22.

31 Ibid., p. 317.

32 Michael Morris, 'The Ethics and Politics of the Caged Layer Hen Debate in New Zealand', *Journal of Agriculture and Environmental Ethics*, CCIX/5 (2006), pp. 495–514.

33 Cited in Joan Dunayer, *Speciesism* (Derwood, MA, 2004), p. 85.

34 Cited in Gilbert White, *The Natural History of Selbourne*, ed. R. M. Lockley (New York, 1949), p. 187.

35 Ulisse Aldrovandi, *Aldrovandi on Chickens*, vol. II, Book XIV [1600], trans. L. R. Lind (Norman, OK, 1963).

36 Chatarina Krångh, personal communication, September 2009.

37 Patrick H. Zimmerman et al., 'Thwarting of Behaviour in Different Contexts and the Gakel-call in the Laying Hen', *Applied Animal Behaviour Science*, LXIX (2000), pp. 255–64.

38 Jonathan Balcombe, *Pleasurable Kingdom* (London, 2006), p. 186.

39 Cited in Temple Grandin and Mark Deesing, 'Distress in Animals: Is it Fear, Pain or Physical Stress?'. See www.grandin.com/welfare/fear.pain.stress.html

40 See www.telegraph.co.uk/news/worldnews/1534177/So-who-are-you-calling-bird-brain-Chatter-of-chickens-proves-they-are-brighter-than-we-thought.html

3 CHICKENLORE

1 Venetia Newall, *An Egg at Easter* (London, 1971), p. 2.

2 Virginia Hamilton, *In the Beginning: Creation Stories from around the World* (New York, 1988), p. 21.

3 Donna Rosenberg, *World Mythology* (Chicago, 1994), p. 361.

4 Barbara C. Sproul, *Primal Myths* (New York, 1979), p. 212.

5 Newall, *An Egg at Easter*, p. 9.

6 Rosenberg, *World Mythology*, p. 404.

7 Edain McCoy, *Ostara: Customs, Spells and Rituals for the Rites of Spring* (St Paul, MN, 2002), p. 10.

8 Newall, *An Egg at Easter*, p. 344.

9 Lewis Carroll, *Through the Looking-Glass* [1871] (Bath, 1962), p. 81.

10 Judika Illes, *Element Encyclopedia of Witchcraft* (London, 2005), p. 193.

11 David Pickering, *Cassell's Dictionary of Superstitions* (London, 1995), p. 175.

12 Jonathan Swift, *Gulliver's Travels* [1726] (Arlington, TX, 1980), p. 56.

13 Steve Roud, *Penguin Guide to the Superstitions of Britain and Ireland* (London, 2003), p. 176.

14 Frederick J. Simoons, *Eat Not This Flesh* (Madison, WI, 1961), p. 68.

15 Illes, *Element Encyclopedia of Witchcraft*, pp. 192–3.

16 Edward Smedley and W. Cooke Taylor, *Occult Sciences* (Whitefish, MT, 2003), p. 333.

17 Pam Percy, *The Complete Chicken* (Stillwater, MN, 2002), p. 26.

18 Boria Sax, *The Mythical Zoo* (Santa Barbara, CA, 2001), p. 67.

19 Simoons, *Eat Not This Flesh*, p. 69.

20 Alan Dundes, *The Cockfight: A Casebook* (Madison, WI, 1994), p. 7.

21 Ibid., pp. 243–4.

22 Ibid., p. 252.

23 Ibid., p. 259.

24 Cited in Ulisse Aldrovandi, *Aldrovandi on Chickens: The Ornithology of Ulisse Aldrovandi*, vol. II, Book XIV [1600], trans. L. R. Lind (Norman, OK, 1963), p. 143.

25 Ibid., pp. 142–3.

26 Simoons, *Eat Not This Flesh*, pp. 69–74.

27 Percy, *The Complete Chicken*, p. 66.

28 Page Smith and Charles Daniel, *The Chicken Book* (Athens, GA, 2000), p. 55.

29 Ibid., p. 25.

30 Percy, *The Complete Chicken*, p. 18.

31 Ibid., p. 66.

32 Francis X. King, *Encyclopedia of Fortune-telling* (London, 1988), p. 173; see also www.cultural-china.com and www.chinainfoonline.com

33 Percy, *The Complete Chicken*, p. 64.

34 Illes, *Element Encyclopedia of Witchcraft*, pp. 50, 386.

35 Percy, *The Complete Chicken*, p. 63.

36 Jeremy Hobson and Celia Lewis, *Keeping Chickens* (Cincinnati, 2007), p. 10.

37 Yvonne Chireau, *Black Magic: Religion and the African American Conjuring Tradition* (Berkeley, CA, 2003), p. 101.

38 Smith and Daniel, *The Chicken Book*, p. 32.

39 Michael Bright, *Beasts of the Field: The Revealing Natural History of Animals in the Bible* (London, 2006), p. 241.

40 A version is in Percy, *The Complete Chicken*, p. 21.

41 Richard E. Strassberg, *A Chinese Bestiary* (Berkeley, CA, 2002), p. 103.

42 Ibid., p. 193.

43 Andreas Johns, *Baba Yaga* (New York, 2004), p. 166.

44 See http://en.wikipedia.org/wiki/Kikimora

45 See www.barnyardcraft.com/kurinyi_bog.htm

46 See www.barnyardcraft.com/liderc_nadaly.htm

47 John Matthews and Caitlin Matthews, *Element Encyclopedia of Magical Creatures* (London, 2005), p. 354.

48 Ibid., p. 245.

49 Javier Ocampo-Lopez, *Mitos y leyendas de Antioquia la Grande* (Barcelona, 2001), pp. 134–5.

50 Carey Miller, *Dictionary of Monsters and Mysterious Beasts* (London, 1987), pp. 11–13.

51 Matthews and Matthews, *Element Encyclopedia of Magical Creatures*, p. 130.

52 Ibid., p. 354.

53 Karl P. N. Shuker, *Extraordinary Animals Worldwide* (London, 1991), pp. 31–3.

4 POPULAR CHICKENS

1 Susan Squier, 'Chicken Auguries', *Configurations*, XIV (2006), pp. 69–86.

2 Charlie Chaplin, dir., *Gold Rush* (United Artists, 1925).

3 Ray Bradbury, 'The Inspired Chicken Motel', in *I Dream the Body Electric* (London, 1971), p. 61.

4 Ibid., pp. 66–7.

5 Jon-Stephen Fink and Mieke van der Linden, *Cluck! The True Story of Chickens in the Cinema* (London, 1981), pp. 73–4.

6 H. G. Wells, *The Food of the Gods* [1904] (London, 2001).

7 Mikhail Bulgakov, *The Fatal Eggs* [1925] (London, 2003).

8 See www.bk.com/en/us/campaigns/subservient-chicken.html

9 See www.casinochicken.com

10 See www.upc-online.org/chicken_flying_contests.html

11 See www.peta.org/about/victoryItem.asp?VictoryID=349

12 See www.miketheheadlesschicken.org

13 Peter Lord and Nick Park, dirs, *Chicken Run* (Aardman Animations, 2000).

14 Mark Dindal, dir., *Chicken Little* (Disney, 2005).

15 Mark McNay, *Fresh* (Edinburgh, 2007), p. 74.

16 Rob Levandoski, *Fresh Eggs* (New York, 2002).

17 Frances O'Roark Dowell, *Chicken Boy* (New York, 2005), pp. 55–6.

18 Clare Druce, *Minny's Dream* (Cambridge, 2004)

19 See www.poultrygeistmovie.com

20 See www.slantmagazine.com/film/film_review.asp?ID=3653

21 See www.guardian.co.uk/film/filmblog/2008/jun/02/poul-trygeistripsaneworific

22 See http://en.wikipedia.org/wiki/Disco_Liesmoby

23 Cited in Pam Percy, *The Complete Chicken* (Stillwater, MN, 2002), p. 73.

24 Carol J. Adams, *The Pornography of Meat* (New York, 2003).

25 Nick Fiddes, *Meat: A Natural Symbol* (London, 1991), p. 95.

26 Alice J. Hovorka, 'The No. 1 Ladies' Poultry Farm: A Feminist Political Ecology of Urban Agriculture in Botswana', *Gender, Place and Culture*, XII/3 (2006), pp. 207–25.

27 Psyche A. Williams-Forson, *Building Houses out of Chicken's Legs: Black Women, Food and Power* (Chapel Hill, NC, 2006), p. 55.

28 Judith Brett, 'The Chook in the Australian Unconscious', *Meanjin*, XLV/1 (1986), pp. 246–9.

29 Fiona Farrell, *Book Book* (Auckland, 2003), pp. 49–63.

5 *GALLUS GRAPHICUS*

1 Pam Percy, *The Complete Chicken* (Stillwater, MN, 2002), p. 69.

2 Ibid., pp. 70–74.

3 See www.languedoc-france.info/06141212_cockerel.htm

4 Percy, *The Complete Chicken*, p. 78.

5 Ibid., p. 76.

6 Jonathan Burt, 'Animals in Visual Art from 1900 to the Present', in *A Cultural History of Animals in the Modern Age*, ed. R. Malamud (Oxford, 2007), p. 167.

7 Percy, *The Complete Chicken*, p. 75.

8 Ibid., p. 78.

9 John Berger, 'Animal World', *New Society*, XVIII (1971), pp. 1042–3.

10 Jonathan Burt, 'The Aesthetics of Livingness', *Antennae*, v (2008), pp. 4–11.

11 Percy, *The Complete Chicken*, pp. 77–8.

12 Ibid., p. 85.

13 James M. Dennis, *Renegade Regionalists* (Madison, WI, 1998), p. 103.

14 Patricia Junker, *John Steuart Curry: Inventing the Middle West* (New York, 1998), pp. 41, 69.

15 Karen Davis, *Review of Ben Austrian, Artist* by Geoffrey D. Austrian, *Poultry Press*, X/2 (2000). See www.upc-online.org/summer2000/ben_review.html

16 Joseph Helfenstein, *Deep Blues: Bill Traylor, 1854–1949* (New Haven, 1999); Susan Mitchell Crawley, *The Life and Art of Jimmy Lee Sudduth* (Montgomery, AL, 2005).

17 Patricia Bjaaland Welch, *Chinese Art: A Guide to Motifs and Visual Imagery* (North Clarendon, VT, 2008), pp. 86–7.

18 See www.britannica.com/EBchecked/topic/298027/Ito-Jakuchu

19 See www.kokodac.com

20 Doug Argue, personal communication, 25 May 2009. See also http://dougargue.com

21 Burt, 'Animals in Visual Art', pp. 176–7.

22 Shannen Hill, 'Iconic Autopsy: Postmodern Portraits of Bantu Stephen Biko', *African Arts* (2005): www.thefreeli brary.com/_/print/PrintArticle.aspx?id=140707295

23 Jovian Parry, *The New Visibility of Slaughter in Popular Gastronomy*, unpublished MA dissertation, University of Canterbury.

24 See www.koenvanmechelen.be/content/view/114/88889030/lang,en/

25 Huang Du, 'Sheng Qi's Body and Discourse'. See www.fourfinger.net/text03.htm

26 Pinar Yolacan, *Perishables* (London, 2005), p. 8.

27 Angela Singer, personal communication, 29 September 2008.

28 Nato Thompson, 'Interview with Nicolas Lampert', in *Becoming Animal: Contemporary Art in the Animal Kingdom* (Cambridge, MA, 2005), pp. 80–81. See also www.machineanimalcollages.com

29 Yvette Watt, personal communication, 24 April 2009.

1 See www.vivausa.org/index.html

2 Jane Dixon, *The Changing Chicken* (Sydney, 2002), p. 83.

3 Ibid., p. 91.

4 Harry. R. Lewis, 'America's Debt to the Hen', *National Geographic* (April 1927), p. 455.

5 Steve Striffler, *Chicken: The Dangerous Transformation of America's Favorite Food* (New Haven, 2005), p. 15.

6 William Boyd, 'Making Meat: Science, Technology and American Poultry Production', *Technology and Culture*, XLII (2001), p. 638.

7 Lewis, 'America's Debt to the Hen', p. 455.

8 Ibid., p. 453.

9 G. W. Wrentmore, *The Battery System of Poultry Keeping* (London, 1931), pp. 7–8, 79 [emphasis in original].

10 Ibid., p. 39.

11 John Steele Gordon, 'The Chicken Story'. See www.american heritage.com/articles/magazine/ah/1996/5/1996_5_52.shtml

12 Carl Weinberg, 'Big Dixie Chicken Goes Global', *Business History Conference* (2004), p. 4.

13 Donald D. Stull and Michael J. Broadway, *Slaughterhouse Blues* (Belmont, CA, 2004), p. 38.

14 Striffler, *Chicken*, p. 7.

15 J. F. Gordy, 'Broilers', in *American Poultry History, 1823–1973* (Mount Morris, NY, 1974), p. 167.

16 Roger Horowtiz, *Putting Meat on the American Table* (Baltimore, MD, 2006), p. 113.

17 Karen Davis, *Prisoned Chickens, Poisoned Eggs* (Summertown, TN, 1996), pp. 85–6.

18 See www.dpichicken.org/faq_facts/docs/ Myths&FactsPosters.pdf

19 Humane Society of the United States (HSUS), 'Broiler Industry Report'. See www.humanesociety.org/assets/pdfs/farm/ welfare_broiler.pdf

20 Gordon, 'The Chicken Story'.
21 Page Smith and Charles Daniel, *The Chicken Book* (Athens, GA, 2000), p. 239.
22 Gordon, 'The Chicken Story'.
23 Striffler, *Chicken*, p. 25.
24 Orlen Grunewald et al., 'Chickens in the Feedlot: The Tyson-IBP Merger' (2009). See www.supereco.com/company/tyson-foods/
25 Stull and Broadway, *Slaughterhouse Blues*, p. 46.
26 Hattie Ellis, *Planet Chicken* (London, 2007), p. 83.
27 See www.kfc.com
28 Eric Schlosser, *Fast Food Nation* (New York, 2001), p. 140.
29 Lois Watson, 'Weight Up What You Eat', *Sunday Star Times* (12 April 2009), p. A7.
30 Michael Greger, in *Mad City Chickens*, dirs Tashai Lovington and Robert Lughai (Tarazod, 2009).
31 Ahang Liying et al., 'Present and Future of China's Broiler Industry', in *Poultry Beyond 2005*, ed. R. J. Diprose et al. (Christchurch, NZ, 2001), p. 25.
32 Eriberto P. Lozada, 'Kentucky Fried Chicken in Beijing', in *The Cultural Politics of Food and Eating*, ed. J. L. Watson and M. L. Caldwell (Oxford, 2004), p. 166.
33 See www.viva.org.uk/campaigns/chickens/broilerfactsheet.htm
34 Michael Watts, 'Afterword: Enclosure', in *Animal Spaces, Beastly Places*, ed. C. Philo and C. Wilbert (London, 2000), p. 299.
35 Cited in Davis, *Prisoned Chickens*, p. 86.
36 Cindy Engel, *Wild Health* (London, 2002), p. 105.
37 Ellis, *Planet Chicken*, p. 41.
38 Ibid., p. 42.
39 HSUS, 'Egg Industry Report'. See www.humanesociety.org/assets/pdfs/farm/welfare_egg.pdf
40 Erik Marcus, *Meat Market* (Boston, MA, 2005), p. 16.
41 A. R. Gerrits, 'Method of Killing Day-Old Chicks Still Under Discussion', *World Poultry*, XI/9 (1995), p. 28.
42 Clare Druce, *Chicken and Egg: Who Pays the Price?* (London, 1989), p. 5.

43 S. Colson et al., 'Motivation to Dust-bathe of Laying Hens Housed in Cages and in Avaries', *Animal*, I (2007), pp. 433–7.

44 HSUS, 'Egg Industry Report'.

45 Catherine Amey, *Clean, Green and Cruelty-Free?* (Wellington, NZ, 2008), p. 47.

46 Davis, *Prisoned Chickens*, p. 55.

47 Hugh Black and Neil Christensen, *Comparative Assessment of Layer Hen Welfare in New Zealand* (Wellington, NZ, 2009), p. 5.

48 C. A. Weeks et al., 'Comparison of the Behaviour of Broiler Chickens in Indoor and Free-Range Environments', *Animal Welfare*, III (1994), pp. 179–92.

49 Karen Davis, 'The Dignity, Beauty and Abuse of Chickens'. See www.upc-online.org/thinking/dignity.html

50 Michael Appleby et al., *Poultry Behaviour and Welfare* (Wallingford, Oxfordshire, 2004), p. 83.

51 Davis, *Prisoned Chickens*, p. 52.

52 Delmarva Poultry Industry, Inc., 'How Broiler Chickens Are Marketed'. See www.dpichicken.org

53 Marcus, *Meat Market*, p. 24.

54 Virgil Butler, 'Whistleblower on the Kill Floor', *Satya* (February 2006). See www.satyamag.com/feb06/butler.html

55 Ellis, *Planet Chicken*, p. 68.

56 Lauren Ornelas, 'Making Chickens Count', *Satya* (February 2006). See www.satyamag.com/feb06/ornelas.html

57 Freeman Boyd, 'Humane Slaughter of Poultry: The Case against the Use of Electrical Stunning Devices', *Journal of Agricultural and Environmental Ethics*, VII/2 (1994), pp. 221–36. Importantly, as this book was going to press, two major chicken meat suppliers in the United States announced that they would be ceasing electric immobilization and moving to controlled atmosphere killing (CAK), a method of slaughter that renders all chickens unconscious before they are even removed from transport crates.

58 Striffler, *Chicken*, p. 94.

59 Butler, 'Whistleblower'.

60 See www.dpichicken.com/index.cfm?content=news&subcontent= details&id=336

61 Michael Greger, *Bird Flu: A Virus of Our Own Making* (New York, 2006), p. 122.

62 Ramona Cristina Ilea, 'Intensive Livestock Farming: Global Trends, Increased Environmental Concerns and Ethical Solutions', *Journal of Agriculture and Environmental Ethics*, XXII (2009), pp. 153–67.

63 E. Lipton, 'Poultry Poses Growing Potomac Hazard', *Washington Post* (1 June 1997).

64 Tom Harkin, 'Animal Waste Pollution in America: An Emerging National Problem'. Report compiled by the Minority Staff of the United States Senate Committee on Agriculture, Nutrition and Forestry for Senator Tom Harkin (December 1997).

65 D. Bell, 'An Egg Industry Perspective', *Poultry Digest* (January 1990), p. 26.

66 M. James, 'Poultry Plant to Pay $6 Million for Polluting', *The Sun* (8 May 1998).

67 'Where Do All the Feathers Go?', *Satya* (February 2006). See www.satyamag.com/feb06/schmidt.html

68 Druce, *Chicken and Egg*, p. 4.

69 Greger, *Bird Flu*, pp. 167–9.

70 Dixon, *The Changing Chicken*, p. 151.

71 Horowitz, *Putting Meat on the American Table*, p. 125.

72 Gerard Kuester, cited in Greger, *Bird Flu*, p. 48.

73 William Boyd, 'Making Meat', p. 663.

74 See http://articles.latimes.com/2004/dec/09/science/ sci-chicken9

EPILOGUE: APPRECIATING CHICKENS

1 Adrian Franklin, *Animals and Modern Cultures* (London, 1999).

2 Tashai Lovington and Robert Lughai, dirs, *Mad City Chickens* (Tarazod, 2009).

3 Mark Lewis, dir., *The Natural History of the Chicken* (PBS, 2000).

4 See www.mixedreality.nus.edu.sg/index.php?option=com_
content&task=view&id=26&Itemid=60

5 Clare Druce, 'Chickens' Lib: The Diary of a Campaign', unpublished
manuscript (London, 2009).

6 See www.openrescue.org

7 See www.kentuckyfriedcruelty.com/pdfs/alicewalker.pdf

8 Philip Armstrong, 'Farming Images', in *Knowing Animals*,
ed. L. Simmons and P. Armstrong (Leiden, 2007), pp. 122–3.

9 See http://www.upc-online.org

10 pattrice jones, 'Roosters, Hawks and Dawgs: Toward an Inclusive,
Embodied Eco/Feminist Psychology', *Feminism and Psychology*,
XX/3 (2010).

11 Aubrey H. Fine, *Handbook on Animal Assisted Therapy* (London,
2010); see also Jessica Hall, 'Chicken Therapy Helps Abused
Children Conquer Fear' (Reuters, 9 February 2000):
http://www.shanmonster.com/chicken/news/news021.html

12 See www.mrjoy.net

Select Bibliography

Adams, Carol, *The Pornography of Meat* (New York, 2003)
—, *The Sexual Politics of Meat* (New York, 2010)
Aldrovandi, Ulisse, *Aldrovandi on Chickens*, vol. II, Book XIV [1600], trans. L. R. Lind (Norman, OK, 1963)
Appleby, M. C., J. A. Mench and B. Hughes, *Poultry Behaviour and Welfare* (Cambridge, 2004)
Coe, Sue, *Dead Meat* (New York, 1996)
Crawford, R. D., *Poultry Breeding and Genetics* (Amsterdam, 1990)
Davis, Karen, 'Thinking Like a Chicken: Farm Animals and the Feminine Connection', in *Women and Animals: Feminist Theoretical Explorations*, ed. Carol J. Adams and Josephine Donovan (Durham, NC, 1995)
—, *The Holocaust and the Henmaid's Tale* (New York, 2005)
—, *Prisoned Chickens, Poisoned Eggs* (Summertown, TN, 1996)
Dixon, Jane, *The Changing Chicken: Chooks, Cooks and Culinary Culture* (Sydney, 2002)
Druce, Clare, *Chicken and Egg: Who Pays the Price?* (London, 1989)
—, *Minny's Dream* (Cambridge, 2004) [a children's story]
Duncan, J. H., and Penny Hawkins, eds, *The Welfare of Domestic Fowl and Other Captive Birds* (Chicago, 2010)
Dundes, Alan, *The Cockfight: A Casebook* (Madison, WI, 1994)
Ekarius, Carol, *Storey's Illustrated Guide to Poultry Breeds* (North Adams, MA, 2007)
Ellis, Hattie, *Planet Chicken* (London, 2007)
Glass, Ira, and Tamara Staples, *The Fairest Fowl: Portraits of*

Championship Chickens (New York, 2001)

Graham, Chris, *Choosing and Keeping Chickens* (London, 2006)

Greger, Michael, *Bird Flu: A Virus of Our Own Making* (New York, 2006)

Grimes, William, *My Fine Feathered Friend* (New York, 2002)

Gurdon, Martin, *Hen and the Art of Chicken Maintenance* (Guilford, CT, 2005)

—, *Travels with My Chicken* (Guildford, CT, 2005)

Haraway, Donna, 'Chicken', in *When Species Meet* (New York, 2007)

Hatkoff, Amy, *The Inner World of Farm Animals* (New York, 2009)

Horowitz, Roger, *Putting Meat on the American Table: Taste, Technology, Transformation* (Baltimore, 2006)

Masson, Jeffrey M., chapter Two in *The Pig Who Sang to the Moon* (New York, 2003)

Newall, Venetia, *An Egg at Easter* (London, 1971)

Percy, Pam, *The Complete Chicken* (Stillwater, MN, 2002)

—, *The Field Guide to Chickens* (Stillwater, MN, 2006)

Potts, Annie, 'With Respect to Chickens', in *Animals and Society: An Introduction to Human–Animal Studies*, ed. M. de Mello (New York, 2011)

Rogers, Lesley, *The Development of Brain and Behaviour in the Chicken* (Wallingford, Oxfordshire, 1995)

Smith, Page, and Charles Daniel, *The Chicken Book* (Athens, GA, 2000)

Striffler, Steve, *Chicken: The Dangerous Transformation of America's Favorite Food* (New Haven, 2005)

Vallortigara, Giorgio, 'The Cognitive Chicken', in *Comparative Cognition*, ed. E. A. Wasserman and T. R. Zendall (Oxford, 2006)

Verhoef-Verhallen, Esther J. J., *The Complete Encyclopedia of Chickens* (Lisse, 2009)

Walker, Alice, 'Why Did the Balinese Chicken Cross the Road?', in *Living By the Word: Selected Writings, 1973–1987* (Phoenix, AZ, 1988), pp. 170–73

Williams-Forson, Psyche A., *Building Houses out of Chicken's Legs: Black Women, Food and Power* (Chapel Hill, NC, 2006)

Associations and Websites

British Hen Welfare Trust
www.bhwt.org.uk

Chicken Run Rescue [chicken adoption and rehabilitation centre]
www.brittonclouse.com/chickenrunrescue

Chocowinity Chicken Sanctuary
www.chocochickensanctuary.org

Farm Sanctuary
www.farmsanctuary.org/issues/factoryfarming/eggs [egg industry]
www.farmsanctuary.org/issues/factoryfarming/poultry [chicken
meat]

United Poultry Concerns [advocacy, education and sanctuary]
www.upc-online.org

Vine Sanctuary
(formerly Eastern Shore Chicken Sanctuary and Education Center)
[rehabilitation for fighting roosters and other chickens]
www.bravebirds.org/projects/education-advocacy

ADVOCACY

Compassion in World Farming
www.ciwf.org.uk

Kentucky Fried Cruelty (run by PETA)
www.kentuckyfriedcruelty.com
Mr Joy [advocacy, education and chicken therapy]
www.mrjoy.net

Open Rescue [chicken rescue operations]
www.openrescue.org

Save Animals from Exploitation [anti-battery farming campaign]
http://safe.org.nz/Campaigns/Battery-Hens

ART

Attention Chicken!
www.machineanimalcollages.com/Pages/Installations/
AttentionChicken.html

Mary Britton Clouse
http://www.mnartists.org/artistHome.do?rid=5365

Graphic Witness [Sue Coe's website]
www.graphicwitness.org/coe/enter.html

What Makes for a Grievable Life? ashley watson (2009)
www.mefeedia.com/watch/27975067

Yvette Watt
www.nzchas.canterbury.ac.nz/associates/watt.shtml

Acknowledgements

In many ways this book has been a collective effort. I have received advice and encouragement from numerous friends and colleagues. My thanks go to Carol J. Adams, Steve Baker, Jonathan Balcombe, Gay Bradshaw, Mary Britton Clouse, Sue Coe, Julie Cupples, Karen Davis, Naina Devi, Clare Druce, Sarah Forgan, Carol Freeman, Ashley Fruno, Nicola Gavey, Ross Gibbs, Carol Gigliotti, Victoria Grace, Donna Haraway, Erin Howard, pattrice jones, Hilda Kean, Marti Kheel, Chatarina Krångh, Brigid Lenihan, Alphonso Lingis, Sharon McFarlane, Susan McHugh, Charmaine McLaren, Patty Mark, Brett Mizelle, Michael Morris, Rebecca Plante, Rio Rossellini, Nigel Rothfels, Karen Saunders, Boria Sax, Julia Schlosser, Tanja Schwalm, Angela Singer, Anna Smith, Susan Squier, Alisha Tomlinson, Shelley Trueman, Sherryl Vint and Yvette Watt. Special thanks to Rebecca Plante for the chicken talisman she gave me right at the start of this project.

I very much appreciate the generosity of those organizations and individuals who donated photographs of chickens or images of art for this book: Farm Sanctuary, People for the Ethical Treatment of Animals, United Poultry Concerns, Save Animals From Exploitation (NZ), Chocowinity Chicken Sanctuary, Eastern Shore Chicken Sanctuary, Open Rescue, Carol J. Adams, Douglas Argue, Andrew Cates, Adrian David Cheok, Mary Britton Clouse, Sue Coe, Larry Z. Daily, Douglas DeVore, Clare Druce, Kay Evans, Davie Hinshaw, Jane Howarth, pattrice jones, Gary Kaplan, Chatarina Krångh, Nicolas Lampert, Mark Lewis, Patty Mark, Amy Pearl, Cappi Phillips, Carol Rhodes-Wittich, Angela Singer, Perry Spiller, James Teh, Alisha Tomlinson, Yvette Watt and Susan Wiren.

I am grateful to Sarah Forgan, Harry Kerr, Sonja Rooney and Rio Rossellini for their wonderful original illustrations; Douglas Horrell for his help with image preparation; the University of Canterbury College of Arts and School of Humanities for research grants to cover image costs; Jonathan Burt, Michael Leaman, Harry Gilonis and Martha Jay for their advice and patience; Sally Borrell for her conscientious assistance in the early days of this project; Jovian Parry for his diligence obtaining images and proof-reading; and Philip Armstrong for his ruthless editing and spot-on suggestions.

Thanks to my extended family for their ongoing support: Jill Vosper and Harry Kerr, Roxane Vosper, Verity Kerr, Ola Roff and the late Syd Roff, Justine Roff, Betty Boniface, Doff and Ian Armstrong. Thanks also to my BFF, Sharon McFarlane, for her sunny distractions.

Chicken is inspired by companion hens over the years. My parents James Mervyn and Faith Katrina Potts responded without fuss to my refusal at a very young age to eat chickens. Living in a rural town surrounded by poultry farms, this was viewed by others as extraordinary lenience on their part. Now I realize how important their acceptance has been to me, and how it has shaped subsequent beliefs about and enjoyment of human–animal relationships, as well as my strong love for chickens.

Philip Armstrong has nurtured me and the 'chooks' through this book. His compassionate approach to life and commitment to challenging speciesism in its many forms continue to be a great inspiration.

This book is dedicated to all chickens born to and killed for agribusiness, scientific research and entertainment; and to those special humans who educate, advocate and provide refuge for these birds.

Photo Acknowledgements

The author and publishers wish to express their thanks to the below sources of illustrative material and/or permission to reproduce it:

Photo Acorn Photo Agency, Perth, Western Australia: p. 116; © ADAGP, Paris and DACS, London 2011: p. 116; photo Ansel Adams/Library of Congress, Washington, DC (Prints and Photographs Division): p. 144; from Ulisse Aldrovandi, *Ornithologiae tomus . . . Liber Decimusquartus qui est de Pulveratricibus Domesticis . . .* (Bologna, 1600): p. 65; image courtesy of Alimdi.net/rgb fotografie: p. 36; photo Alinari/Rex Features: p. 108; photo courtesy Nir Alon: p. 80; courtesy of the artist (Doug Argue) and Doug Argue Studios, San Francisco: p. 126; photos courtesy of the author: pp. 11, 25, 50, 127, 174; British Museum, London (photos © The Trustees of the British Museum): pp. 14, 58, 66, 85; Carnegie-Stout Public Library, Dubuque, Iowa: p. 120; image courtesy of Andrew Cates and the John Leech Archive: p. 20; collection of the artist (Judy Chicago) and needleworkers (image courtesy of Judy Chicago and Through the Flower): p. 148; © the artist (Mary Britton Clouse): pp. 53, 132; image courtesy of the artist (Sue Coe): p. 126; image courtesy of Larry Z. Daily: p. 110; image courtesy of Karen Davis and United Poultry Concerns: p. 162; photo Jerry Daws/Rex Features: p. 171; photos © DaydreamsGirl/2011 iStock International Inc.: pp. 24 (right), 28; photo © DeannaBean/2011 iStock International Inc.: p. 15; photo © DreamBigPhotos/2011 iStock International Inc.: p. 63; image courtesy of Clare Druce: p. 178; from Daniel Giraud Elliot, *A Monograph of the Phasianidae (Family of Pheasants)* (New York, 1872): p. 10; image courtesy of Kay Evans: p. 156; images courtesy of Farm Sanctuary: pp. 154,

Index